The Law of Driverless Cars: An Introduction

Alex Glassbrook

Barrister

Temple Garden Chambers

Law Brief Publishing

Published 2017 by
Law Brief Publishing
30 The Parks
Minehead
Somerset
TA24 8BT

www.lawbriefpublishing.com

Paperback: 978-1-911035-28-2
Kindle ebook: 978-1-911035-42-8

To Roger, who sat on seatbelts,

and to Isaac and Theo, who might not have to.

Contents

INTRODUCTION

"*All vehicles have their defects*" wrote Lord Justice Fletcher Moulton in 1909:

> "*Those that are drawn by horses are liable to cause or suffer accidents through the horse becoming frightened, or slipping and falling. If they are four-wheeled vehicles, such as horse-drawn omnibuses, they are liable to swerve and skid on greasy pavements. Self-propelled vehicles have the advantage that there are no horses that can be frightened, or can fall. Such vehicles are also under more complete mechanical control, and can ordinarily be pulled up in a much shorter distance... There is nothing which points to any of them being so unsuitable for use in street traffic as to constitute a nuisance...*"[1]

The new technology of "self-propelled vehicles" must have seemed alarming to many, over a century ago. What followed – particularly the increases in speed as motor technology advanced – fuelled the fears.

Measures to restrict speed were adopted (including the notorious "man with a flag", walking ahead of the chugging car) and speed limits altered – up and down – as technology improved and motor vehicle use increased. After much debate and many casualties, compulsory insurance for drivers was introduced, followed by voluntary agreements (by an insurance industry under threat of further legislation) to provide safety nets of insurance for those injured by uninsured drivers.

Thus, law and motor insurance grew together.

A "self-propelled vehicle", frightening automaton though it would have seemed, was still under human control. At the outset, that human control might not have been very effective and the machinery unresponsive to fine control. But – as Mr Justice Fletcher Moulton indicated – the car was still easier to control than a horse.

---

1 *Wing v London General Omnibus Company* [1909] 2 K.B. 652, 667

And the invention did things that vehicles (carts, horse or steam-driven vehicles) could not easily have done. Quick to start, the car became a means of recreation, and a car-ride a treat in itself. Grand new roads were built (and to the bewilderment of later generations), a trip along a motorway became a leisurely pastime.

The motor car came to represent wider ideas – no longer a means merely to get from one place to another, it transformed the journey into its own experience. Radio added a further possibility – the car as hub of entertainment.

So the use of cars increased – as did private ownership. The law expanded (regulation, the criminal and civil laws – the latter largely concerned with the liability of drivers to those injured by their driving). And so did insurance (towards compulsory coverage for injury to third parties).

Now, we are in another stage of the evolution. The mutations that the car is undergoing are more profound than many of us realise (in terms of the technology and the effects upon our environment) and they are gathering pace very quickly indeed. Some of the central innovations are of a different character to the "driver assist" technologies (cruise and distance control, parking assist systems, for example) to which we have become accustomed.

The mutations will – hopefully – include engines powered by less environmentally-damaging means than fossil fuels. Electric cars will not meet that need unless the means of energy-generation are also environmentally friendly. Energy-generation is a pressing subject, though not for this book.

(This book deals specifically with "road" vehicles. So the intriguing world of automated air traffic – automated drones – is not its focus.)

(I also have not dealt with environmental law implications – for example those relating to the manufacture of such vehicles, including battery manufacture, nor with the land-use and planning implications of reduced numbers of road vehicles. Those are subjects for another work – perhaps for a future edition.)

The key mutation in road vehicles, for present purposes, is the development of "driverless technology". That is to say, the production, for use on public roads, of a car that can, without human intervention, operate and control its own speed and direction.

As at the time of writing, there is lively debate as to whether a truly "driverless" car will ever be produced. The questions posed include: whether any machine can be perfectly safe; whether it would be ethical to give a machine the capacity to decide the least bad of several injurious outcomes, if any outcome were likely to result at best in serious injury and at worst in death (the "kill decision", as it has been called). A car manufacturer has announced its intention to include driver controls such as a steering wheel even on its driverless cars. The Department of Transport of the British government has just announced its intention to extend the compulsory insurance regime to include collision claims involving a semi-autonomous car driven in automated mode. Whether this will be the final shape of things remains to be seen.

For the purpose of this book, I make several assumptions.

First, I assume that the driverless car (in the sense of a car that can, without human intervention, operate and control its own speed and direction) will be produced and that many such cars will eventually be in use – perhaps within decades. That is speculative and perhaps even controversial, but the march of technology suggests that it is a likely outcome.

Secondly, I assume that driverless technology will never achieve perfection – whether in its safety or in any other aspect of its operation – but that it will achieve a standard of safety high enough for full driverless use on public roads. Perfection appears an unrealistic goal. There is debate as to the measure of safety that should be required of driverless technology, but a consensus appears to be forming that it should, at least, be demonstrably far less prone to error than a human driver. Some favour a threshold of at least twice as safe as a human driver, others a much higher threshold – eg. to 99.99% certainty of safety.

Thirdly, I assume that driverless technology will at some time be a safer technology than human driving. That is not to denigrate any driver (myself included). It is a reflection of the human susceptibility to all sorts of distractions when operating a machine as fast and numerous in its features as a car, and of the constant potential for collisions when many humans drive, inevitably distracted, together.

If those assumptions are correct, then it is likely that driverless cars will come into use on British roads and that they will significantly reduce the number of road traffic accident casualties currently occurring.

So, back to driverless car technology today.

Distinct from the "Driver Assistance" devices with which we are already familiar (cruise control, parking assist, etc), perhaps the most striking of the developments in driverless technology is the growth of "machine learning" – a type of artificial intelligence that not merely recognises symbols programmed into it and reacts (as a factory robot might), in set, programmed ways. Instead, the fully driverless car might be a machine that *learns*: developing and displaying intelligence that its programmers did not design and might not even have foreseen[2].

The car of the future might be one that takes its own decisions. It might be privately owned, or it might be an "app car", summoned to your door from a tap on your smartphone. It will do a lot of things that vehicles currently do either not at all, or only to a relatively primitive extent.

The greatest (and perhaps most startling) of those new features will be the car's ability to gather, use and store information. Technology challenges our attitudes to privacy daily. Driverless cars will accelerate that

2 See Hod Lipson and Melba Kurman's account of an experimental artificial intelligence network spontaneously learning to respond to human faces, at page 232 of "Driverless: Intelligent Cars and the Road Ahead" (Lipson and Kurman, MIT Press, Cambridge, Massachusetts, 2016)

process. The process of gathering data for driverless cars has already begun (by, for example, the mapping of "street views").

But intelligent cars will not only follow existing maps. As they drive, they will draw further details onto the map, as well as learn about the area in more detail (eg. its vehicle and human population). Intelligent cars will talk to each other. Their conversations and conclusions will be sent to databases – corporate and public. The value of that data (in, variously, financial terms to the developers and personal terms to users) will be considerable.

Challenges to the use of that data by those harvesting it – whether by operating the system legitimately or by hacking into any part of it criminally – are bound to arise.

So data security and privacy might become new battlegrounds of risk, where law and insurance might even find themselves most often engaged.

That is all to look forward, to exciting future technologies. But throughout this high-tech landscape, old technology is bound to persist. Just as cars in town and countryside still make way for the horse and the bicycle, so will driverless cars travel a world in which humans (riders, cyclists, pedestrians) and entirely non-electronic objects (including wholly manual cars) appear. The driverless car will not be able to communicate with everything.

The old and the new will forever co-exist. And, in the early years of the new technology, there will be a transitional period. That will be especially legally complex, not least because of the unfamiliarity of the new technology and of its implications, and the clash with existing, non-automated motor vehicles.

There might soon be visible changes. The reduction in the number of drivers and – as I shall go on to discuss – the likely reduction in the number of privately-owned vehicles if app cars take hold (and as fuel-driven cars become more expensive to maintain) – will change even the appearance of our streets, as the use of space for car parking diminishes.

The arrival of driverless cars will require new laws, as well as old laws deployed in new ways. How we relate to vehicles from now on – and what they do for and to us – will be very different from our current experience.

Driverless cars have the potential to offer individual, easily-accessible transport to people who have not qualified – or have not been allowed to qualify, for reasons of health or perceived ability – as drivers of vehicles. It is a potentially liberating technology.

We are on the cusp of this technology, so much of what follows is unavoidably speculative. It is impossible at this stage to explore every legal byway down which driverless cars may steer. In the interests of brevity I have not followed every imaginative route. I have tried to limit myself to the brief of providing an "Introduction" to the emerging law.

But there are many reasons to be inquisitive, and to look forward to the imminent, exciting changes in our everyday surroundings.

The changes are likely to be both dramatic and – if harnessed properly – highly beneficial. That depends to a great extent upon getting the law right. Rights and responsibilities need to be re-ordered fairly and effectively. Maintaining a status quo will not be the solution for a changed landscape.

Across the Atlantic, a decade after Lord Justice Fletcher Moulton's judgment and in a different context, Justice Oliver Wendell Holmes described the foundation of the law of free speech under the U.S. Constitution as "the free trade in ideas". He portrayed society deciding its laws by discussion, as best it could, requiring it "every year, if not every day, to wager our salvation upon some prophecy based upon imperfect knowledge"[3].

3 The dissenting opinion of Justice Oliver Wendell Holmes (Justice Brandeis concurring) in *Abrams v United States* (1919) 250 U.S. 616, which concerned the criminality (or not, in the dissenting justices'view) of pro-Bolshevik pamphlets published in the U.S. during the First World War and after the Russian revolution. Justice Holmes described the theory of free speech underlying the U.S. Constitution as "free trade in ideas ... the power of the thought to get itself accepted in the competition of the market" (630).

Justice Holmes did not believe that legal debate was only for the lawyers. The present discussion, as to how the law should be framed to driverless cars, is active, mainly online. It spans several countries and numerous fields of interest.

As it continues, that discussion will surely involve (to name but a few, in no order of priority) robotics engineers, artificial intelligence specialists, road and motor engineers, planners, road safety groups, architects, schools and their students, policymakers and all users of streets and vehicles (which is to say everyone). Inventors, manufacturers, government, lawyers and insurers are not the only groups with an interest in this debate.

This book is a small contribution to that discussion. It is a fascinating, optimistic topic.

Presuming (heavily) to borrow a phrase from Oliver Wendell Holmes, this book is an exercise in everyday prophecy. And it comes against the wider background of a national legal migration, away from a pan-national system of law that is the present source of rules in the UK as to (amongst other subjects) motor insurance and data protection. Those will be core areas in future transport law.

The effects of Brexit will be profound, but the route (shortly to be the subject of a first government White Paper, in the wake of today's judgments of the Supreme Court in the Gina Miller case[4]) is so far unmapped and unknown. That journey will be as important for the modern law of transport (to borrow the title of the government's imminent bill) as it will be for many other areas of everyday life in Britain.

---

4 *R (on the application of Miller and another) v Secretary of State for Exiting the European Union* [2017] UKSC 5

I make no special pleading for any errors in this book – prophetic or otherwise – which are all my own.

But then the law is, to quote Justice Holmes, "*an experiment, just as all life is an experiment*"[5]. So that is some reassurance.

Alex Glassbrook
Temple Garden Chambers, London
24th January 2017

5 *Abrams v U.S.* (above).

HOW TO USE THIS BOOK

This book deals with the law of England and Wales. Its objectives are to:

- describe the situations that are likely to arise from driverless technology
- set out the legal principles that are likely to apply, and to
- offer a view of the future legal landscape, to all those who need to consider the legal consequences of the new technology.

The overall purpose is to do so in as concise and accessible a way as possible. But it is an exciting subject, so I hope the reader will forgive signs of enthusiasm.

The writer is a practising barrister with 20 years' experience as an advocate and legal adviser in cases involving insurance, personal injury, consumer and damages law – often against a road traffic background – including the defence of suspected fraudulent claims. He is also a driver and cyclist, with a lively interest in policy relating to transport and technology.

This is, to the publishers' knowledge, the first book published in the United Kingdom on the law of driverless vehicles.

You can read the book entirely, OR:

If you would like a brief overview of the topic, read "PART ONE: THE CHANGING LANDSCAPE" (Chapters 1 to 3), which describes (Ch.1) the current law, (Ch.2) the long-term future law and (Ch.3) the transitional interim phase of the law of driverless cars.

If, having read Part One, you would like more detail about any of the topics raised, then read any of the chapters that interest you in:

PART TWO: THE NEAR FUTURE (which deals with the interim phase, while driverless technology attempts to establish itself)

And

PART THREE: THE FULLY-AUTOMATED FUTURE (which deals with possible law, in the event that the driverless vehicle becomes the dominant type of road vehicle). This is necessarily the part of the book in which imagination takes flight, though I have tried to keep it to the route set by the earlier chapters.

PART ONE

THE CHANGING LANDSCAPE

CHAPTER ONE
CURRENT VEHICLES, CURRENT LAW

INTRODUCTION

Motor vehicles are now almost always under the control of a human driver when in motion. The current exceptions are devices like cruise control and parking assist, which temporarily take control of one or more functions of the vehicle. But the driver must be able to take control of the vehicle while those functions are in use.

Disputes arise mainly from collisions involving vehicles, though there are other litigious situations.

Insurance for road vehicles became compulsory when parliament made it so, because of the quantity of road traffic accidents causing injuries. Drivers are still required to be insured against third party injury and losses. The motor insurance industry also provides cover, on a voluntary basis through the Motor Insurer's Bureau, for third parties injured by uninsured and untraced drivers.

Several types of prototype driverless vehicles are being tested in some parts of the UK. Insurance for prototype vehicles has been issued.

TOPICS IN THIS CHAPTER

This chapter sets out:

- the main legal obligations in the UK relating to road traffic and to prototype driverless vehicles being tested in the UK

- the most common legal disputes.

- the current insurance regime for road traffic, in outline

OBJECTIVES OF THIS CHAPTER

The objectives of this chapter are to familiarise the reader with:

- the current litigation landscape of road traffic, under human control
- the current insurance regime, and
- the law as it relates to prototype driverless vehicles now being tested on British roads

and to set the context for the discussion of likely changes in the law, to accommodate driverless vehicles (the subject of Chapter 2).

1. MAIN LEGAL OBLIGATIONS

I deal only with the main legal obligations relating to driven vehicles (to set the scene for considering, in the following chapter, the likely alterations in relation to driverless cars). What follows is not an exhaustive list of legal obligations in relation to cars. For such information I would refer the reader to more detailed works[1].

I set out some of the terms to categorise the legal obligations:

- "International law" refers to rules agreed between countries including the UK.
- "Regulatory law" describes rules made by the British state (often by secondary legislation – a statutory instrument – made in accordance with powers provided by statute, ie. an Act of parliament; sometimes directly by statute). For example: regulations in

1 Eg. Bingham & Berrymans' Personal Injury and Motor Claims Cases (Lexis Nexis, 14th edition, 2014), The Law of Motor Insurance (Robert Merkin, Jeremy Stuart-Smith, Sweet & Maxwell, 2nd edition, 2015), Encyclopedia of Road Traffic Law and Practice (Robert Ward, updated periodically, Sweet & Maxwell).

relation to the structural requirements for vehicles. The insurer's statutory duty (directly under the Road Traffic Act 1988) to satisfy a judgment obtained against a driver – the "compulsory insurance" provisions, leading to the description of the insurer as "Road Traffic Act insurer" – I also place into this category.

- "Criminal law" to describe conduct which parliament has prescribed as punishable by criminal penalty (including endorsement of driving licence, fine, imprisonment), eg. speeding or careless or dangerous driving.
- "Private law" to describe the relationships between private individuals, which would include the relationship between a manufacturer and a purchaser of a car. Private law "remedies" would include, eg. the ability of a customer to sue a manufacturer for damages for breach of contract, if she was sold a defective vehicle.

DRIVING

International Law

Standards for vehicles are set primarily by international agreement.

The United Nations Convention on Road Traffic 1968 (The Vienna Convention) contains detailed rules concerning the design of road vehicles. In the UK, the Road Vehicles (Construction and Use) Regulations 1986, made under section 41 of the Road Traffic Act 1988 and enacting EU law within the UK, also deal with the detailed requirements of vehicles.

The main principle, which runs through the Convention (subject to some exemptions – set out below) is that a moving vehicle must always be under the control of a driver who is both capable and qualified to control it. That principle is of obvious relevance to the prospect of "driverless" vehicles.

Article 8 provides that every moving vehicle or combination of vehicles shall have a driver, who shall:

- possess the necessary physical and mental ability to drive
- be in a fit physical and mental condition to drive
- have (unless he is a learner driver practising driving) the knowledge and skill necessary for driving the vehicle
- at all times be able to control his vehicle

In April 2014, Article 8 was amended as follows:

> *"5bis. Vehicle systems which influence the way vehicles are driven shall be deemed to be in conformity with paragraph 5 of this Article and with paragraph 1 of Article 13, when they are in conformity with the conditions of construction, fitting and utilization according to international legal instruments concerning wheeled vehicles, equipment and parts which can be fitted and/or be used on wheeled vehicles"*

– which is to say that automated systems are permitted, if they comply with the 1968 Convention. But if they do not:

> *"Vehicle systems which influence the way vehicles are driven and are not in conformity with the aforementioned conditions of construction, fitting and utilization, shall be deemed to be in conformity with paragraph 5 of this Article and with paragraph 1 of Article 13, when such systems can be overridden or switched off by the driver."*

So, the amended version of the 1968 Convention allows autonomy, but only if there is always a driver in the car who can "override or switch off" the autonomous systems (note the plural – all of those systems). So the 2014 amendment foresees a switch that turns the car (presumably instantly) back into a driven car.

Article 13 of the 1968 Convention provides that:

> *"Every driver of a vehicle shall in all circumstances have his vehicle under control so as to be able to exercise due and proper care and to be at all times in a position to perform all manoeuvres required of him. He shall, when adjusting the speed of his vehicle, pay constant regard to the circumstances ..."*

Article 41 provides that the driver must hold a permit (ie. be qualified to drive, having passed a test set by a competent public authority).

Annex 5 ("Technical conditions concerning motor vehicles and trailers") provides that:

> *"46. Every motor vehicle shall be equipped with a strong steering mechanism which will allow the driver to change the direction of the vehicle, easily, quickly and surely"*

Annex 5 also provides a list of exemptions to the Technical requirements for vehicles:

> *"60. For domestic purposes, Contracting Parties may grant exemptions from the provisions of this Annex in respect of:*
>
> > *(a) Motor vehicles and trailers which, by virtue of their design, cannot exceed a speed of 30 km (19 miles) per hour on a level road of whose speed is limited by domestic legislation to 30 km per hour;*
> >
> > *(b) Invalid carriages, i.e. small motor vehicles specially designed and constructed - and not merely adapted - for use by a person suffering from some physical defect or disability and normally used by that person only;*
> >
> > *(c) Vehicles used for experiments whose purpose is to keep up with technical progress and improve road safety;*
> >
> > *(d) Vehicles of a special form or type, or which are used for particular purposes under special conditions;*

(e) Vehicles adapted for use by handicapped persons."

It is clear (from the 2014 amendment) that none of those exemptions (from the original 1968 convention) remove the need for a driver, because the 2014 amendment expressly provides that "*Vehicle systems which influence the way vehicles are driven*", ie. any sort of automated system, must be capable of being "*overridden or switched off by the driver*".

The UK signed the Vienna Convention but has not ratified it, so is not legally bound by it. It is, however, obliged not to act in a way that would defeat or undermine the treaty's objective and purpose. The Vienna Convention describes the desire of the contracting parties "to facilitate international road traffic and to increase road safety through the adoption of uniform traffic rules".

It is arguable that the introduction of driverless cars to the UK, without prior agreement of uniform rules relating to such cars (and, consequently, wholesale reform of the Vienna Convention to remove or amend the "driver" requirement) might put the UK in a position of undermining the Vienna Convention, in which case the Convention would need to be amended to allow fully driverless cars to be used in the UK.

Regulation 104 of the Road Vehicles (Construction and Use) Regulations 1986 makes the legal requirement for a human driver in the UK clear:

> *"No person shall drive or cause or permit any other person to drive, a motor vehicle on a road if he is in such a position that he cannot have proper control of the vehicle or have a full view of the road and traffic ahead."*

The current position, therefore, appears to be that vehicles can, at most, be only partially autonomous. A qualified driver must be able at any time to deactivate the autonomous driving system and take control of the car – by steering and bringing it safely to a halt.

The decision of a certain car manufacturer outside the UK to retain steering wheels in its autonomous vehicles therefore follows the Vienna Convention.

The British government takes the position that the need for a further change in the Convention to allow automated vehicles on the roads is "arguable" and that the Convention "*is not considered an obstacle in the UK. The UK has signed but not ratified the convention and testing is consistent with proper driver control*"[2]

It is not clear whether the UK would resist the Convention solely on the basis of its non-ratification, should the technology reach such a high standard that fully driverless cars, ie. lacking the need for a qualified driver to override the system, become available.

The likelihood is that such technology (and the consequent redundancy of the human driver) would lead to the amendment of the Convention and of the British Construction & Use Regulations. That might be some time away.

Testing of Prototype Autonomous Vehicles in the UK

The British government's position is that existing regulation allows testing of prototype vehicles on public roads in the UK, "providing a test driver is present and takes responsibility for the safe operation of the vehicle"[3].

That is consistent with the Vienna Convention. It does not deal with the possibility of "driverless" vehicles, however. At some stage – if and when the technology reaches that stage – that question will need to be addressed.

---

2 "The Pathway to Driverless Cars: Summary report and action plan" (UK Department of Transport, February 2015) p.20, para.3.4.

3 Dept of Transport Feb 2015 paper (above) at page 20, para.4.6.

For the time-being, the human driver remains very active. Which brings us to:

The Domestic law of driving

Drivers owe obligations to:

- the individuals around them (passengers, other drivers, pedestrians and others) to take reasonable care for their safety and
- society at large, not to contravene the criminal laws relating to driving

Those obligations are carried into effect by the civil law of negligence (which allows for awards of compensation against a driver for any injury or loss proved to have been caused by his negligent driving) and by the criminal law (which provides for criminal sentences – including fines and imprisonment, for driving contravening the standards of the criminal law).

Criminal Offences

To give just a few examples, it is an offence to:

- drive carelessly
- drive dangerously
- drive at a speed exceeding the legal limit
- drive whilst under the influence of drink or drugs
- use a car on a public road without a policy of insurance in force

All of those offences exist because a human driver controls a car and because loss of control is likely to involve significant threat to the safety of others.

A culpable loss of control is therefore deemed so serious as to constitute criminal behaviour – even if the culpability is careless rather than intentional. A failure to insure (ie. the driver's failure to insure others against the consequences of his carelessness) is also a criminal offence.

The standard to which the prosecution must prove that the driver of a car committed all the elements of a criminal offence is "beyond a reasonable doubt", ie. to make the magistrate or jury (whichever might be the finder of fact) sure, beyond a reasonable doubt, that the offence was committed. "Beyond a reasonable doubt" is termed "the criminal standard of proof".

Private Law Obligations on Driver – civil law

The civil law provides a remedy in damages (compensation) for any personal injury or damage to property caused by a driver's failure to take reasonable care for the safety of anyone reasonably foreseeably affected by such failure (which amounts, in practical terms, to anyone in his vicinity – passengers, pedestrians et al).

Such failure to take reasonable care is termed "negligence".

Unless the driver admits negligence, the injured party must prove the driver's negligence (as well as the fact and extent of any damage, and that it was caused by the negligence) on "the balance of probabilities", which is known as "the civil standard of proof".

The "balance of probabilities" means that a party (the claimant) must prove his case to more than 50% of likelihood. If the allegedly negligent driver's (the defendant's) case is just as likely as the claimant's, the claim will fail, "for want of proof", because the balance of probabilities is not tilted in the claimant's favour.

However, if the claimant persuades the court (a District Judge or Circuit Judge of the County Court, or a judge of the High Court in more serious cases) that his case is more likely than the defendant's case – even just 1% more likely – then the claim will succeed. For that reason, the "balance of probabilities" standard of proof is expressed as being a likelihood of at least 51%.

Obligations of Third Parties to the driver (pedestrians, cyclists, etc)

The term "Third Party" is distinct from the parties to the insurance contract: the insured driver (first party) and motor insurer (second party).

Third parties also owe duties to the driver.

A list of all possible third parties would be too long for present purposes. The most common third parties in an RTA case are:

- the driver of the other vehicle involved (or motorcyclist) or
- an injured non-driver (pedestrian, cyclist etc)

Drivers and motorcyclists have the same duties in common: those prescribed by penalties under the criminal law and the private law duty (chiefly in negligence) owed to those reasonably foreseeably affected by their acts or omissions.

Cyclists are subject to similar criminal offences (careless and dangerous cycling, cycling under the influence of drink or drugs – plus cycling on the pavement). They are also subject to the civil law duties, chiefly the duty of care in negligence

Contributory Negligence

An allegation of negligence is often countered by an allegation of contributory negligence, ie. that the injured party also failed to take care of himself.

The effect of such a finding is to reduce the injured party's damages by a percentage representing the extent of his fault and its causative effect.

Obligations of Manufacturer of Vehicle

Manufacturers might be held liable, under civil law, in damages for any defects in their vehicle:

- to the purchaser, under the law of contract, and
- to others affected by the defect (eg. passengers or other third parties injured by such defect) in the law of negligence or product liability.

These causes of action will be considered in further detail as we consider the features of driverless cars.

Obligations of Repairers of Vehicles

Repairers will owe duties pursuant to the law of contract, to provide a service in accordance with the terms of the contract. A defective service will expose them to an action for damages.

Obligations of Commercial Operators using vehicles – Employers, Providers of Public Transport etc

Employers and public transport companies owe duties of care to those using their vehicles – whether their employees (to whom the employer owes duties of care including the duty to provide safe work equipment) or passengers (to whom duties are owed just as a passenger in a private car is owed a duty – save that the duty arises not only in negligence but also in contract in the paying passenger's case).

2. MOST COMMON DISPUTES

The landscape is dominated by the law of negligence, which governs cases involving road traffic collisions. Other areas of law can apply (eg. the law of nuisance, if, for example, a load falls from a vehicle and obstructs the highway). But the law of negligence dominates.

The following types of claim dominate the road traffic litigation landscape now:

road traffic collisions (damages actions, for personal injuries, property damage and/or accident management charges such as hire costs)

Insurance disputes (eg. as to coverage where there is a collision, including fraudulent claims against insurers and compensation under the Motor Insurers Bureau scheme)

Those are the main features. Other civil law disputes that might arise from road traffic include:

consumer disputes (eg. product quality or liability disputes between customers and manufacturers, services disputes between customers and repairers, consumer credit disputes between consumers and financiers)

and

Disputes as to the state of the highway (eg. actions under the Highway Act, nuisance by eg. obstruction of the highway)

Of the various types of claim, road traffic collision claims dominate. These result from the unreliability of human decision-making when driving at speed. Consequently, such cases are fought almost entirely in the law of negligence – the law of carelessness.

3. CURRENT MOTOR INSURANCE

Human decision-making when driving has been proved to be so unreliable and dangerous as to persuade lawmakers to impose a compulsory system of insurance, supported by (under the threat of legislation) voluntary agreements between motor insurers to compensate those injured by uninsured as well as untraced drivers[4].

So, in the UK, motor insurance has evolved to provide a sequence of safety nets, on the assumption that the negligently-injured "third party" wishing to claim damages will (provided he complies with rules as to issuing his claim or notifying the MIB in time) be compensated by one of the several insurance schemes. Hence "compulsory third party insurance".

There are economies and advantages to such a system – both for insurers and for injured third parties. At the time of writing (though possibly not post-Brexit[5]), injured motorists can elect to sue the motor insurer of a negligent driver, instead of the driver himself[6]. And (a possible casualty of Brexit) that right to sue can cross European borders, allowing European drivers to be sued in the EU state in which the injured person lives[7].

---

4 See the Motor Insurers Bureau Uninsured Drivers Agreements and the Untraced Drivers Agreement, at https://www.mib.org.uk

5 Britain's exit from the European Union as a result of the referendum of 23 June 2016: in process at the time of writing.

6 See the European Communities (Rights Against Insurers) Regulations 2002, SI 2002/3061

7 An *Odenbreit* action, under the Court of Justice of the European Union (CJEU)'s judgment in *Schadeverzekeringen NV v Odenbreit* (Case C-463/06) [2007] ECR I-11321

A motor insurer, even if not sued directly in its own name under the 2002 EC Regulations, has and usually exercises a contractual right to conduct the defence of a claim made against its insured driver, on his behalf, because it bears the financial risk of the defence failing. The court recognises the insurer's financial interest in the claim, eg. by the principle of subrogation (allowing the insurer to claim its expenditure in the insured's name) and by the insurer's right to be added as a defendant to a claim, should it wish, to guard its financial interest[8].

Court procedure recognises the insurer's role. For example, the court protocols, by which the court encourages and may enforce efficient pre-claim conduct by the parties, include a protocol specific to low value personal injury claims in road traffic accidents. That protocol provides a form that must be sent to the defendant's motor insurer[9]. The claim then moves through closely-defined stages. The recovery of fixed amounts of costs is provided for by set formulae.

And the system of costs in personal injury claims is now subject to a system proposed by insurers as an economic response to the elevation of inter-party costs orders in injury cases by lucrative "no win, no fee" agreements (now outlawed).

Under the current system motor insurers pay the "base" (meaning, in this sense, the un-enhanced) costs of almost all claims, irrespective of the outcome of a case, unless a few narrow exceptions (eg. the insurer had previously offered a better deal than the injured party obtained at trial, or that the claim is in some sense improper, eg. by fundamental dishonesty).

That system (known by the technical label "qualified one-way costs shifting", or "QoCs") was seen by insurers as an acceptable price for the extinction of inflated "uplifts" of costs for winning claimants in "no win, no fee" agreement personal injury cases. Those uplifts had often

---

8 *Gurtner v Circuit* [1968] 2 QB 587

9 See the Pre-Action Protocol for Low Value Personal Injury Claims in Road Traffic Accidents.

resulted in the doubling of the bill to defendant's insurers for lawyers' time costs.

So the present system of compensation for road traffic injuries – and thereby the law of road traffic – is underpinned and shaped by insurance.

The compulsory nature of motor insurance is both a source of revenue for insurers and a source of considerable expenditure – not least due to the level of exposure to fraudulent collision claims for personal injury compensation.

SUMMARY

The law and the litigation landscape have been shaped entirely by the human driver's control of cars. That is the source of almost all law and of the major part of claims made in relation to road vehicles.

If driverless cars arrive on our roads, they will alter the landscape in almost every foreseeable aspect. We shall glimpse that new world in the next chapter.

CHAPTER TWO
FUTURE VEHICLES, FUTURE LAW

INTRODUCTION

The objective of this chapter is to introduce the reader to a distant, strange but exciting future. This world will be fleshed out in more detail in the chapters in part three of the book ("The fully automated future").

Imagine a future in which machines can learn. The owner of a new car, taking delivery of it, selects "set up" on its touchscreen, then steps out of the car and watches it set out on its first drive around the neighbourhood.

It is learning about these streets, its occupants, the cats and bicycles to which it needs to pay attention. After half an hour it returns, loaded with the extra information it needs most often (in addition to the factory-loaded street view maps and intelligence it already has).

The car has looked at the area itself. But it has also made contact with all of the other cars in the area, and has learnt (on a server in the cloud) all that they have to tell it. It is ready to take the owner wherever he needs to go. It will continue learning on every journey it takes, and will feed that information into the cloud – to all of the other cars.

That is a possible vision of the future. It is not the only possible version. It is not even the whole story – there are many implications. For example, private ownership of cars might cease to be as much a feature of life in the UK as it is now. People might decide that ownership of a car is an expense needed only in certain circumstances (eg. for those who need personal mobility so frequently as to justify the expense), or during certain periods of life.

Many people might decide to use "app cars" instead – driverless cars that arrive at their location at the touch of a symbol on their smartphone or watch. Sharing journeys in "Communal app cars", which can plan a journey more efficiently, synchronising with the journeys of fellow passengers, would cut fares.

This future might not be very distant: sudden advances in artificial intelligence and battery technologies, for example, could deliver intelligent, electric vehicles (and the obsolescence of fossil fuel, driven vehicles) with surprising rapidity. The arrival of the internet and of touchscreen and smartphone technologies provide examples of this.

This vision (like other versions of the driverless future) is of transport that performs the function of taking us from one point to another in a very different way from a driven motor vehicle.

Importantly, it is a vision of transport that itself uses information: a car that interprets (in present terms) its own Sat-Nav directions and learns: not only on every journey but constantly, as it updates from the cloud, learning the experiences of all other driverless cars.

It might even be the case that driverless cars will depend upon *data* as much as they depend upon a means of propulsion. And that factor – implying the exchange of very large amounts of data, constantly – will be a particular challenge to society and to lawmakers.

That will be so especially in relation to our understanding of privacy, which is already challenged by advancing technology.

OBJECTIVES OF THIS CHAPTER

The objectives of this chapter are to examine:

- The changes that the introduction of mainly robotic cars might bring about to a future society
- The consequences of those changes
- The risks and disputes that might arise, and

- The ways in which the law might respond.

TOPICS IN THIS CHAPTER

This chapter sets out:

- A version of the UK in the future, in which driven cars have been superseded by fully driverless cars with deep (or "machine learning") artificial intelligence. It explores the implications of that future – eg. for car ownership, for the arrangement of streets and for transport as a business.

It also describes, in outline, the topics to be dealt with in more detail in the chapters in part three ("The fully automated future"):

- The disputes that might arise in that future
- The laws that would apply – both the established laws and those that might develop with the technology, and
- The nature of risks and therefore of insurance, in that world.

The objective of this chapter is to familiarise the reader with a strange new future world, in which the functions of cars – and the legal landscape – have changed radically.

(Chapter Three – and Part Two of this book – will deal with the interim period, between our present situation and the Fully Autonomous Future. I describe this intervening period as "The Untidy Transition". For the moment, I shall outline a version of the Fully Autonomous Future.)

1. IMPLICATIONS OF A DRIVERLESS FUTURE

I have already set out a part of the vision of a fully autonomous future – the fully intelligent, fully independent, driverless vehicle. The owner needs to keep it charged but, in essence, it will probably behave like a tablet computer or smartphone: it will tell the owner when it needs to be charged, when its software needs to be updated and when there is a problem.

The latter point – in the case of a problem occurring while the car is in motion – will presumably be accompanied by the car halting itself in a safe position. The car will have similar distance and speed limiting features to those even now available on driven cars – except those features would be much more sophisticated. In particular, they will also speak to other driverless vehicles. So the fully driverless car should, in theory, be able to halt reasonably safely, in a way co-ordinated with any other driverless vehicles around it, even if a glitch occurs while in motion.

So the fully driverless car should be reliable – if not perfectly so, then to a very high standard indeed (as close to the unattainable 100% as could realistically be expected). Defects and accidents should be very rare – certainly much rarer than the road traffic accidents in the world of driven vehicles.

The "kill question" (the runaway car, for example, faced with the dilemma of probably killing one of two people, so having to decide which life is worthier) should occur extremely rarely, if at all. The other features of the driverless car (which would include, for example, speed limitation) ought significantly to reduce the risk of that question arising.

Furthermore, rather like a smartphone compared to an earlier telephone, the driverless car will perform many more functions than those necessary for its central purpose of providing transport. These will include, for example, providing entertainment via an online music and visual library, as well as access to email, social media etc. This will be how the passenger (formerly the driver) occupies his or her journey time.

This is all, unavoidably, speculative. But we can make informed guesses as to the implications of fully driverless technology. The broad implications might be as follows:

- Cars "for hire" will be much more accessible. Even today, taxis are more accessible than previously, chiefly because of smartphone technology.

- The evolution of taxis might lead towards driverless "app cars", ie. driverless vehicles called to a location at short notice via a smart mobile device (eg. smartphone or watch).

- As availability of "app cars" improves, the advantages of owning a car privately might diminish. The burdens of private ownership (eg. the need retain space for a vehicle outside a home, the need to pay privately for the electricity necessary to charge the vehicle, servicing charges, diminution in value of a vehicle) might in some situations (eg. in a city, where space is scarce) become outweighed by the advantages of app cars.

- Private and public spaces previously set aside for parking privately owned cars might then become free for other uses. Cars would take up less space generally.

- Road safety will, in all likelihood, be highly improved. Maximum speeds would be pre-programmed into the driverless cars. Driverless cars will, in any event, drive at a speed appropriate to the surroundings. Driverless cars will not be impatient, nor feel threatened whilst driving, nor drive aggressively. The frequency of accidents and the case for compulsory motor insurance will diminish.

Some of those implications have economic consequences (eg. for car manufacturers, who might need to manufacture vehicles in greater proportions for businesses than for individual owners; for local authorities and central government who would be less able to rely upon parking fees, fines and road tax as a means of revenue; for motor insurers, if the law requiring compulsory insurance were repealed).

Some have social consequences (eg. the liberation of parking space for other uses; need for lesser space for road traffic when planning the design of towns; the increase of pedestrian and cycling traffic in spaces less susceptible to road traffic accidents; reduction of pollution, with the reduction of vehicles).

There are many beneficial points. There will also be negative consequences. For example, the burden on energy production would increase with the introduction of recharged electric, multi-functional vehicles, unless vehicle numbers were reduced by a lesser share of privately owned vehicles. Job losses are likely to be suffered in the driving sector.

In relation to data, the benefits of cars accessing data are likely to be countered by some significant challenges:

- Driverless cars will have access to large quantities of information, much of which might be described as personal (eg. location data) and some of which will be confidential (eg. email).

- Driverless cars will use some of that information (eg. location data, data as to numbers of people in certain locations) in order to carry out their central purpose of providing safe transport.

- Commercial and public organisations providing driverless vehicles or services connected with them will have access and legal obligations in relation to that data.

And there will still be accidents. There is no realistic prospect of traffic ever becoming entirely robotic. People will still walk and ride (bicycles, electrical motorbikes and horses). Indeed, a benefit of the fully driverless future would be the increase in space and safety to do so.

And, for the pleasure of it, some drivers will still want to drive cars (though no doubt subject to heavy tax and required to insure themselves extensively and at considerable cost if doing so on public roads whose motor traffic consists mainly of artificially intelligent, speed-limited, unemotional cars). Human driving might then become a very expensive and closely regulated activity – the equivalent of flying an antique aeroplane today.

So, even in a "driverless" world, perfection will not be attainable. Accidents will occasionally happen. Machinery – however sophisticated – will still go wrong.

What disputes might then arise?

2. DISPUTES IN A DRIVERLESS FUTURE

It is difficult to extract ourselves from the present, so the word "accident" forces itself to the top of any mental list of disputes relating to road traffic.

But the reduced likelihood of accidents, in a world in which self-propelled vehicles would almost all be under robotic control, is very likely to demote accidents to a lower position in the list of risks.

This is an exercise in speculation. But, with that qualification, the risks attached to a driverless road traffic environment might fall into the following categories:

- The risks to passengers (as everybody travelling in a self-propelled vehicle would be). Some passengers would be customers, some would be owners and some would be in the vehicle because their work required of it of them.

- The risks to third party bystanders, ie. those outside the vehicle but close enough to be affected by it.

- The risks to those with a commercial relationship to the vehicle. That could include a business providing the vehicle as part of a service (eg. personal transport or haulage of goods).

- The risk to those responsible for the functioning of the vehicle. That could include the manufacturer or supplier of the vehicle (through their potential liabilities to the purchaser of the vehicle and to the public at large, in relation to any product defects). It could also include the repairer of the vehicle, or the supplier of any software for the vehicle – or indeed of systems outside the vehicle (eg. in the cloud) but which contribute to its functioning.

The damage triggering any of those risks could be in any of the following forms:

- Financial loss

- Damage to reputation

- Physical damage to property

- Personal injury to a passenger or third party

And that damage could come about in a number of ways. For example:

- Financial loss could be the result of equipment (the car) not functioning as well as had been expected or promised. For example, deliveries might not be performed in time. Financial loss might also result from the release of information (whether accidental or as the result of theft of information). A business might suffer financial loss through injury to one of its employees. Or it might suffer loss because it is using the vehicle for its business purposes, so might be found to be strictly, vicariously liable for any damage caused by the vehicle.

- Damage to reputation would be another of the possible results of the release of information, especially confidential information such as might be contained in emails.

- Physical damage to property might result from a malfunction in the vehicle, for example one leading to a collision.

- Personal injury to a passenger or third party might result from a collision. Or it might take the form of psychiatric injury resulting from the release of information.

3. LAW IN A DRIVERLESS FUTURE

On that basis, the present laws that might be engaged in traffic disputes in the more distant, driverless future would include:

- information law, particularly the law of data protection

- the criminal law – which might need revision to include (eg.) data offences specific to driverless vehicles, as well as offences affording protection from crime targeted at such vehicles

- the torts (civil wrongs) of breach or misuse of confidential information and breach of personal privacy

- the law of negligence

- contract law

- the law of product liability

- consumer law

- employers' liability (eg. for work equipment)

- strict and vicarious liability (on the part of a business, if a driverless car were regarded as a robotic driver, and deemed to be an employee)

- the law of damages (including the law relating to contribution and apportionment of liability between several parties)

- insurance law (which might, for reasons considered below, eventually be recast from its present compulsory model)

- regulation of vehicles – particularly the rewriting of the Vienna Convention and of domestic regulations to allow vehicles to be used in fully driverless mode and by those unqualified to or physically incapable of taking over driving

- equality and discrimination law – to provide for such passengers' rights, if they are not provided for in revised regulations

That is probably not a full list, as it is not possible to speculate as to every permutation. I have excluded the laws that are likely to be relevant to the development stage of driverless technology (eg. intellectual property law), as a result of commercial competition between the many businesses involved.

Experience shows that crime is likely to innovate with the technology. Whether it takes the form of hacking for confidential information or even the virtual hijacking of driverless vehicles cannot be predicted with any confidence. But it is bound to look for ways into the new technology.

It is a busy list. And it is very likely that new laws will develop in order to accommodate the new technology. Just as the Road Traffic Acts evolved (and continue to do so) to provide protection through criminal and civil law – mainly through the imposition of a system of compulsory insurance – so will a new generation of acts legislate for the new landscape.

4. RISKS IN A DRIVERLESS FUTURE

So the risks will not arise purely – or even mainly - from vehicle collisions. They will arise from defects, from the carriage of information, from the nature of the tasks for which driverless vehicles are used.

Those tasks will be commercial and might be on a large scale (eg. using "platoons" of co-ordinated, driverless heavy goods vehicles), as well as on the small scale of the "app car".

The proportions in which people relate to cars owner, hirer or passenger – will alter. We might all become passengers not only of a robotic driver but also of taxis. We might decide not to own cars but to call for them at will. "Pay as you go" might become the most common relationship with the cars of the future.

Whatever the relationship turns out to be, it will be governed by different laws – some old, some new, but all in a different shape.

The risks will vary in much the same pattern. There will be significant risk, even in a driverless world at a mature stage of development. Insurers might cover for information-related risks and financial and reputational losses, in addition to the traditional area of third party cover for collision-related injury and loss. The current model, of compulsory insurance because of the injury risk posed by human drivers, is unlikely to survive into the long term.

So that is a picture of a possible world, in the mature stage of driverless vehicles.

That would come after the messy, transitional phase – to which I shall turn in the next chapter.

SUMMARY

Driverless technology is very likely to bring about many benefits, including a marked improvement in road safety and the release of a great deal of public and private space.

However, disputes and legal issues will also arise.

Perhaps the most significant change in the fully driverless future will result from the vehicle's storing of information, which presents particular liability risks, as well as a further test to our already challenged notions of personal privacy. The new landscape will test the law.

CHAPTER THREE
THE UNTIDY TRANSITION

INTRODUCTION

This chapter deals with the interim period, between our present situation and a (possibly) fully driverless future.

This interim period lies in the near-future, when driverless technology has advanced to the point at which experimental "driverless" cars can perceive and react to their surroundings in a significantly safer way than a human driver, so that government legislates explicitly to permit their use on all public roads.

In that near-future phase (I would speculate), manufacturers and signatories to the Vienna Convention will have agreed that, notwithstanding the success of the technology in trials and in now-widespread use on public roads, the requirement that a fit and able driver must be able to take control of the vehicle instantly should remain law and practice.

So, in this scenario, the Vienna Convention remains in its 2014 version, and the UK abides by that version. (I base that assumption upon the 2014 amendment to the Convention and upon the decision by a major manufacturer to retain steering wheels when developing its autonomous vehicles, which appears to follow the Convention). So there must be a human driver in every vehicle, able to take control of the vehicle if circumstances require.

The early phase of driverless vehicles will be "untidy" (my word) because the new, high-tech vehicles will be visitors to an old-fashioned world: one populated by vehicles subject to the control of emotional, distracted humans.

And there will be pedestrians, cyclists, children, roadworks – a wealth of obstacles and problems, in a world not designed for robot vehicles at all.

The visitors will need to adapt to those onerous conditions, while relying upon systems which will still be in the earliest stage of use in the real world, so in some respects untested.

TOPICS IN THIS CHAPTER

This chapter:

- Outlines many of the legal disputes that might arise in that near-future phase, in which "driverless" cars (though not in the fully driverless sense) are in use in the UK, alongside conventional traffic and

- Outlines the laws that would govern those disputes.

OBJECTIVE OF THIS CHAPTER

The objective of this chapter is to outline the laws that will apply in the early landscape of driverless vehicles. The details of those laws will be considered in the chapters in Part Two of the book.

1. LEGAL DISPUTES IN THE "UNTIDY" EARLY PHASE

As stated in the Introduction, there are areas too large for this book (eg. the environmental law implications of driverless technology). I shall not deal with those.

The legal disputes in the early stages of driverless technology being used on the roads can be described in several ways.

There is the hazard of producing a list of technical legal terms, which would be of little general use. At the other extreme, there is also the danger of relegating major, real concerns in the attempt to produce a cosmetically attractive, apparently straightforward list. I will probably commit both faults.

With that apology, my list of the legal areas in which risks are likely to arise, during the early driverless phase, is as follows:

- Data
- Public safety
- Employee safety
- Crime
- Identifying liable parties
- Regulation, and
- Insurance

2. LAWS APPLYING IN THE TRANSITIONAL PHASE

In outline (for more detailed discussion in Part Two), the areas of law likely to be engaged are as follows.

Data

The law as to security of data presently comes mainly from 2 sources:

- The Data Protection Act 1998 (DPA), which is the domestic enactment of European Union law (Directive 95/46/EC) pursuant to the European Communities Act 1972. So that might – by the time we reach our notional near-future – have been repealed, following Brexit.

- The torts (civil wrongs) of breach of confidence, and of misuse of private information[1].

And, as from May 2018, the DPA will be replaced by the European General Data Protection Regulation (the "GDPR"). That will apply even stricter requirements than the DPA. The Information Commissioner reports that, notwithstanding Brexit, the British government will not seek to avoid the commencement of the GDPR in the UK in May 2018 because the UK will then still be a member state of the EU[2]. Whether the GDPR becomes the more permanent form of British data protection law after Brexit remains to be seen.

Data law will be considered in later chapters. Its particular interest, to a discussion of a developing technology, is the rapidity of its development over recent decades. As Lord Justice Keene said of the tort of breach of confidence, in 2001, in *Douglas v Hello! Limited*:

> "... *breach of confidence is a developing area of the law, the boundaries of which are not immutable but may change to reflect changes in society, technology and business practice.*"[3]

Public safety

The chief sources of law guarding public safety in the road traffic field are:

---

1 The latter recognised as a tort by the Court of Appeal in *Google v Vidal-Hall and others* [2015] EWCA Civ 311.

2 "The GDPR will apply in the UK from 25 May 2018. The government has confirmed that the UK's decision to leave the EU will not affect the commencement of the GDPR": https://ico.org.uk/for-organisations/data-protection-reform/overview-of-the-gdpr/introduction/

3 [2001] QB 967, 1011e

- The criminal law relating to driving.
- The civil law of negligence (by the disincentive of damages).
- The civil law relating to product quality (contract law for the purchaser, product liability law for the broader public).
- Domestic regulatory law as to the safety features of vehicles (informed by the Vienna Convention).
- Domestic statutory and regulatory law as to employee safety (dealt with as a separate topic below).
- Insurance law (which provides for compulsory third party insurance for injury and property damage, plus schemes to cover for uninsured and untraced drivers).

Issues that are likely to arise in the laws protecting public safety might include:

- The circumstances in which the driver of a nearly-driverless car (whom I will call "the standby driver", that being the essential characteristic under the Vienna Convention) is required – whether by the criminal law or the law of negligence – to act in the interests of their own and third parties' safety.
- What sort of problem in a nearly-driverless car would constitute a "defect" so as to trigger potential liability for a manufacturer (for example) under product liability or contract law?
- Whether the state should regulate or legislate, during this stage, for specific "nearly-driverless" car situations (the UK government's current position is that, in effect, it will await the situation then regulate, if necessary, in "waves of reform"[4]).

---

4 The latter "waves of reform" approach has been, and remains as at the time time of writing, the preferred governmental approach. See the Department of Transport's

- How the courts or the legislature would apply employer's liability laws in relation to equipment to driverless cars and whether such duties would be strict liability duties or not.

- How a manufacturer would respond to a claim in which the mechanics of the accident could not be explained.

- Whether businesses' vicarious liability for their employees to injured members of the public would extend to strict, vicarious liability for an unexplained defect in a driverless car.

- Whether the compulsory insurance model now applied to road traffic (which the UK government intends to maintain in relation to autonomous vehicles[5]) would remain the permanent model for road traffic including robotic, near-driverless vehicles.

- If not, what other insurance models would emerge? Would manufacturers (for example) take over insurance in relation to some risks (eg. product defects) and offer that with purchase of the car?

Employee safety

The sources of law as to employee safety are:

- The common law (including the employer's duties to provide a safe place of work and safe work equipment).

---

proposals paper, "Pathway to Driverless Cars" July 2016, page 8, and its response paper to the consultation of interested parties, January 2017.

5 in the proposed Modern Transport Bill, due to be published in early 2017. See the January 2017 response paper, cited in the footnote above. The insurance proposal is to continue compulsory third party cover and thereby to allow for a claim against the insurer of a vehicle being driven in automated mode, subject to that insurer's right to seek contribution or indemnity from another (eg. manufacturer) liable for any fault in the automated feature.

- Statutory and regulatory duties (some of which are sourced from European Union law, so are of questionable permanence post-Brexit; liability to compensate in damages, which formerly existed for some of the domestic, regulatory duties of care, was removed in most cases by statute in 2013[6]).

Crime

The criminal offences in relation to driving were set out in chapter 1. Those include offences for a driver (a human driver) to

- drive carelessly
- drive dangerously
- drive at a speed exceeding the legal limit
- drive whilst under the influence of drink or drugs
- use a car on a public road without a policy of insurance in force.

The British government's approach to criminal law reform to accommodate autonomous vehicles, is to leave current laws in place, "taking a step-by-step approach, and regulating in waves of reform" so as to be able "to learn important lessons from real life experiences of driving increasingly automated vehicles".[7]

It seems very likely that new crimes will evolve with the technology.

---

6 Section 69 of the Enterprise and Regulatory Reform Act 2013

7 Department of Transport paper July 2016, page 8, reiterated in January 2017.

Identifying liable parties

Identifying a party liable to criminal or civil penalty has not previously been difficult in road traffic law: it almost always is the driver or (if applicable, in civil law) the business vicariously responsible for him.

Vehicle defect or a defect in the highway, if it arose, might arguably deflect liability elsewhere. It can be raised as a defence by the allegedly liable driver, but is not often proven in current road traffic accident cases.

In the case of an autonomous vehicle, however, the likely prospect in most early accident claims is that the conduct of the automated vehicle will be called into question.

That will engage:

- product liability law (in relation to whether or not the vehicle was defective) and
- the law of negligence (as to whether or not the human driver ought reasonably to have intervened, whether by doing so he would on the balance of probability have prevented or just mitigated the effects of the accident)

In relation to driver intervention, the question will arise as to whether the courts would attach all or only some liability to the driver. Might the courts instead attach a percentage of liability to the manufacturer of a culpable vehicle even if the driver could have intervened to prevent the accident?

Regulation

As appears from the quotations above, the British government has taken a light touch, preferring to see how the technology evolves than prescribing a legislative framework beforehand.

Insurance

The government has so far elected to retain the compulsory insurance model.

Whilst this is a sensible approach in the transitional phase of the technology, however, this will might not be the best model for the fully-driverless future. I shall discuss that point in Part Three.

SUMMARY

The arrival of the first driverless (or nearly driverless) vehicles as a feature on all British roads will drop the deck of legal cards onto the table. How the law might play out is unpredictable, but it is bound to be an untidy picture for some time.

PART TWO

THE NEAR-FUTURE

CHAPTER FOUR
DATA

INTRODUCTION

Too often our mobile telephones ring, showing an unfamiliar number. We pick up, there is an odd silence, then a recorded voice congratulates us on our luck, then goes into the first phrase of a sales pitch as we hang up.

Personal data is already a commodity, of significant commercial value. Data about us has increased, because technology allows data to be gathered and stored with increasing ease. And storage is a problem, because it is insecure.

So far, I have spoken of the need for battery and artificial intelligence technology to improve, before driverless cars will become a reality. But data security technology will have to adapt as well. Because our data is already lucrative and its storage vulnerable.

So vulnerable, in fact, that in May 2018 new European Union data rules will come into effect[1], backed by large fines and applying to any organisation handling the data of European Union citizens.

That has been a reaction to failures in data security in the United States (followed by the 'Safe Harbor' ruling by the Court of Justice of the European Union, the CJEU).

The world of the near-future will be one in which fear of data insecurity will affect individuals and states. This might even prove to be the major risk of what has been termed "the internet of things": the evolution of

1 The General Data Protection Regulation ("GDPR"), due to come into effect within the European Union – and with some effects beyond the EU in 2018. "The GDPR will apply in the UK from 25 May 2018. The government has confirmed that the UK's decision to leave the EU will not affect the commencement of the GDPR" : https://ico.org.uk/for-organisations/data-protection-reform/overview-of-the-gdpr/introduction/

the internet to allow everyday objects to connect and communicate over networks.

Driverless cars might be the most significant of those everyday objects to become conveyors of their users' personal information. If driverless cars become widespread, the transit of our personal data will increase at a very high rate.

The digital age is too far advanced to be held back. But might the early years of driverless cars be those in which we face a difficult decision: either to enact increasingly strict legal protections for our privacy as individuals, or to decide that new services justify individual sacrifices of privacy? To what extent will we agree to give up our privacy? Or might we look increasingly to the courts, for data protection?

OBJECTIVES OF THIS CHAPTER

The objective of this chapter is to familiarise the reader, in summary terms, with the current laws relating to security of data and personal privacy, in the near-future context of driverless vehicles.

CONTENTS OF THIS CHAPTER

This chapter sets out, in the context of driverless vehicles:

- The key laws affecting personal privacy and data security
- How those are in a state of flux and
- How developments in the near-future might affect the risks arising from the use of data by driverless vehicles

1. THE LAWS OF PERSONAL PRIVACY

British law has, relatively recently, developed remedies for misuse of personal information.

That development has been through statutory law enacted by parliament and through the common law (the law interpreted by judges in the courts). International and European Union law have been influential.

The torts (civil wrongs) of breach of confidence and misuse of private information have grown in tandem with statutory law, chiefly two statutes enacted in 1998: the Human Rights Act (which put the rights to privacy under Article 8 and Freedom of Expression under Article 10 of the European Convention on Human Rights (ECHR) on a statutory footing) and the Data Protection Act.

This chapter is concerned largely with the Data Protection Act (DPA). That is the current statute, though it is due to be overtaken by the GDPR in May 2018.

Like the Human Rights Act (HRA), the DPA provides a remedy in compensation for unlawful release of personal information. Unlike the HRA, it is highly specific in its provisions relating to information. Being a creature of international (that is to say, European Union) law, it is arguably more attuned to the character of data in a digital age: borderless and ever-changing.

So it is to the DPA that we should look first, when considering the impact of driverless cars. In particular, the data passing through driverless "app cars", should those come into widespread use (eg. in city centres), will be regulated most by data protection laws.

THE LAWS OF DATA PROTECTION

The law protecting data has, largely because of its borderless nature, been designed at the international level. The current British statute –

the Data Protection Act 1998 – is a creature of European law. It springs from the 1995 Data Protection Directive (95/46/EC).

Under the European Communities Act 1972 (due to be repealed by the UK's withdrawal from the EU), directives of the EU are directly applicable in British law, consistently with the Treaties establishing the EU. At the time of writing, EU law still applies, though that will change with withdrawal from the EU.

Beneath the original 1995 EU Data Directive (95/46/EC) are a number of more specific European laws, all of which have influenced British law. Particularly important European laws include:

- The 2002 "e-Privacy" Directive (2002/58/EC), or "ePD". and
- (applicable from May 2018) the 2016 General Data Protection Regulation (Regulation 2016/679), or "GDPR".

In the present context of protection of data arising from the use of vehicles the key features of European law are:

- The fundamental purposes of EU law to protect the right to confidentiality of communications and data, via the Charter of Fundamental Rights of the European Union, given force by the 2009 Lisbon Treaty. These are often referred to as the "third generation" fundamental rights protected by EU law.
- The aim for legal certainty as to the laws applicable to data protection – this is set out in the foundational 1995 Data Directive

And

- The fundamental requirement for the consent of users of a service to the processing of their data. This is already a guiding principle, enshrined (as described below) in the British Data Protection Act, for example. But its scope is changing, particularly in relation to location data. The ePD emphasises the requirement for users of location data to consent (in all but a few cases) before information is collected[2]. And the EU's data protection working party, reporting in July 2016, recommended that the GDPR should maintain the high level of protection already provided in EU law by the ePD, to safeguard confidentiality.[3]

The British government has already indicated that it intends to bring the GDPR into British law at its commencement in May 2018, despite the Brexit process. The complexity of rewriting data law (especially against a short deadline) might be a motivator, but the international nature of data protection law is probably also a factor in that reasoning.

The Data Protection Act (DPA) 1998

The following is a summary of the effect of the DPA 1998. It does not provide a comprehensive account of the workings of the Act (for such an account the reader is referred to longer, specialist works). It focuses upon the areas of the Act most likely to affect driverless cars (and not upon more remote areas of data protection, eg. exemption for national security).

British data protection law deals with the control and processing of personal data. These terms are broad, but they are defined by section 1 of the DPA 1998 as follows:

"*Processing*" means "*obtaining, recording or holding the information or data or carrying out any operation or set of operations on the information or data, including -*

---

2 See Opinion 03/16 of the Data Protection Working Party (an independent European advisory body on data protection and privacy) on the evaluation and review of the ePD, 19 July 2016, page 4, at http://ec.europa.eu/justice/data-protection/article-29/documentation/opinion-recommendation/files/2016/wp240_en.pdf

3 EU working group on ePD, July 2016, page 4.

(a) organisation, adaptation or alteration of the information or data,

(b) retrieval, consultation or use of the information or data,

(c) disclosure of the information or data by transmission, dissemination or otherwise making available, or

(d) alignment, combination, blocking, erasure or destruction of the information or data"

"Personal data". There are two parts to this:

a. "*Data*" "*means information which—*

(a) is being processed by means of equipment operating automatically in response to instructions given for that purpose,

(b) is recorded with the intention that it should be processed by means of such equipment,

(c) is recorded as part of a relevant filing system[4] *or with the intention that it should form part of a relevant filing system, . . .*

(d) does not fall within paragraph (a), (b) or (c) but forms part of an accessible record as defined by section 68 [records of public health, education, housing and social services authorities – in England and Wales]; or

(e) is recorded information held by a public authority and does not fall within any of paragraphs (a) to (d);

---

4 "Any set of information … structured … in such a way that specific information relating to a particular individual is readily accessible": Section 1(1) DPA

and

b. "*Personal data*" means "*data which relate to a living individual who can be identified—*

> *(a) from those data, or*
>
> *(b) from those data and other information which is in the possession of, or is likely to come into the possession of, the data controller,*

> *and includes any expression of opinion about the individual and any indication of the intentions of the data controller or any other person in respect of the individual*"

To whom the Act applies

The Act imposes obligations upon Data Controllers and Data Processors. It gives rights to Data Subjects.

A Data Controller is "*a person who (either alone or jointly or in common with other persons) determines the purposes for which and the manner in which any personal data are, or are to be, processed*", or is appointed as such under an Act.

A Data Processor "*in relation to personal data, means any person (other than an employee of the data controller) who processes the data on behalf of the data controller*".

A Data Subject is "*an individual who is the subject of personal data*".

A Data Subject has rights, including:

Under Section 7, to:

"be informed by any data controller whether personal data of which that individual is the data subject are being processed by or on behalf of that data controller"

have *that information*, and any information available to the data controller as to the *source* of those data, communicated to him "*in intelligible form*"

Where the data is has constituted or is likely to constitute the sole basis for any decision significantly affecting him, to be informed by the data controller of the logic involved in that decision-taking.

Under Section 11, to prevent the processing of information relating to him for the purpose of direct marketing.

Under Section 12, to request that no decision that significantly affects him is taken solely by automated processing of information.

Under Section 14, to ask the court to order the data controller to rectify, block, erase or destroy inaccurate information.

Under Section 13, to recover compensation for damage and for distress[5] caused by contraventions of the Act by a data controller.

Territorially, Data Controllers fall under the Act if (under Section 5):

"(a) the data controller is established in the United Kingdom and the data are processed in the context of that establishment, or

(b) the data controller is established neither in the United Kingdom nor in any other EEA State but uses equipment in the United Kingdom for processing the data otherwise than for the purposes of transit through the United Kingdom"

---

5 See the *Google v Vidal-Hall* case ([2015] EWCA Civ 311), in relation to damages for distress under Section 13, considered alongside the development of the torts.

A Data Controller falling under (b) must nominate a representative in the UK.

A Data Controller must be registered as such with the Office of the Information Commissioner in the UK (section 17) and notify the OIC of any changes in particulars of registration (section 20). It is an offence to process data without registration and to fail to notify the OIC of such changes (section 21).

Section 55 sets out the criminal offences of unlawfully obtaining or procuring the disclosure of data:

> *"(1) A person must not [save in certain circumstances, eg. prevention of crime or by order of a court] knowingly or recklessly, without the consent of the data controller—*
>
> *(a) obtain or disclose personal data or the information contained in personal data, or*
>
> *(b) procure the disclosure to another person of the information contained in personal data."*

The Data Protection Principles

Section 4 and Schedule 1 of the DPA set out eight "Data Protection Principles":

"SCHEDULE 1
THE DATA PROTECTION PRINCIPLES

Section 4(1) and (2)

Part I The Principles

1 Personal data shall be processed fairly and lawfully and, in particular, shall not be processed unless—

(a) at least one of the conditions in Schedule 2 is met, and

(b) in the case of sensitive personal data, at least one of the conditions in Schedule 3 is also met.

2 Personal data shall be obtained only for one or more specified and lawful purposes, and shall not be further processed in any manner incompatible with that purpose or those purposes.

3 Personal data shall be adequate, relevant and not excessive in relation to the purpose or purposes for which they are processed.

4 Personal data shall be accurate and, where necessary, kept up to date.

5 Personal data processed for any purpose or purposes shall not be kept for longer than is necessary for that purpose or those purposes.

6 Personal data shall be processed in accordance with the rights of data subjects under this Act.

7 Appropriate technical and organisational measures shall be taken against unauthorised or unlawful processing of personal data and against accidental loss or destruction of, or damage to, personal data.

8 Personal data shall not be transferred to a country or territory outside the European Economic Area unless that country or territory ensures an adequate level of protection for the rights and freedoms of data subjects in relation to the processing of personal data."

The conditions for the processing of personal data, under part (a) of the first principle, are as follows:

"SCHEDULE 2
CONDITIONS RELEVANT FOR PURPOSES OF THE FIRST PRINCIPLE: PROCESSING OF ANY PERSONAL DATA

Section 4(3)

1 The data subject has given his consent to the processing.

2 The processing is necessary—

(a) for the performance of a contract to which the data subject is a party, or

(b) for the taking of steps at the request of the data subject with a view to entering into a contract.

3 The processing is necessary for compliance with any legal obligation to which the data controller is subject, other than an obligation imposed by contract.

4 The processing is necessary in order to protect the vital interests of the data subject.

5 The processing is necessary—

(a) for the administration of justice,

(aa) for the exercise of any functions of either House of Parliament,

(b) for the exercise of any functions conferred on any person by or under any enactment,

(c) for the exercise of any functions of the Crown, a Minister of the Crown or a government department, or

(d) for the exercise of any other functions of a public nature exercised in the public interest by any person.

6 (1) The processing is necessary for the purposes of legitimate interests pursued by the data controller or by the third party or parties to whom the data are disclosed, except where the processing is unwarranted in any particular case by reason of prejudice to the rights and freedoms or legitimate interests of the data subject.

..."

Section 4(3) of the DPA sets more demanding conditions for the processing of sensitive data. "Sensitive personal data" is defined by Section 2 as:

"*... personal data consisting of information as to—*

(a) the racial or ethnic origin of the data subject,

(b) his political opinions,

(c) his religious beliefs or other beliefs of a similar nature,

(d) whether he is a member of a trade union (within the meaning of the Trade Union and Labour Relations (Consolidation) Act 1992,

(e) his physical or mental health or condition,

(f) his sexual life,

(g) the commission or alleged commission by him of any offence, or

(h) any proceedings for any offence committed or alleged to have been committed by him, the disposal of such proceedings or the sentence of any court in such proceedings."

Part (b) of the First Data Protection Principle is that sensitive personal data must not be processed unless one of the conditions in Schedule 3 is also met. Schedule 3 provides:

"SCHEDULE 3
CONDITIONS RELEVANT FOR PURPOSES OF THE FIRST PRINCIPLE: PROCESSING OF SENSITIVE PERSONAL DATA

Section 4(3)

1 The data subject has given his explicit consent to the processing of the personal data.

2 (1) The processing is necessary for the purposes of exercising or performing any right or obligation which is conferred or imposed by law on the data controller in connection with employment.

. . .

3 The processing is necessary—

(a) in order to protect the vital interests of the data subject or another person, in a case where—

(i) consent cannot be given by or on behalf of the data subject, or

(ii) the data controller cannot reasonably be expected to obtain the consent of the data subject, or

(b) in order to protect the vital interests of another person, in a case where consent by or on behalf of the data subject has been unreasonably withheld.

4 The processing—

(a) is carried out in the course of its legitimate activities by any body or association which—

(i) is not established or conducted for profit, and

(ii) exists for political, philosophical religious or trade-union purposes,

(b) is carried out with appropriate safeguards for the rights and freedoms of data subjects,

(c) relates only to individuals who either are members of the body or association or have regular contact with it in connection with its purposes, and

(d) does not involve disclosure of the personal data to a third party without the consent of the data subject.

5 The information contained in the personal data has been made public as a result of steps deliberately taken by the data subject.

6 The processing—

(a) is necessary for the purpose of, or in connection with, any legal proceedings (including prospective legal proceedings),

(b) is necessary for the purpose of obtaining legal advice, or

(c) is otherwise necessary for the purposes of establishing, exercising or defending legal rights.

7 (1) The processing is necessary—

(a) for the administration of justice,

(aa) for the exercise of any functions of either House of Parliament,

(b) for the exercise of any functions conferred on any person by or under an enactment, or

(c) for the exercise of any functions of the Crown, a Minister of the Crown or a government department.

(2) The Secretary of State may by order—

(a) exclude the application of sub-paragraph (1) in such cases as may be specified, or

(b) provide that, in such cases as may be specified, the condition in subparagraph (1) is not to be regarded as satisfied unless such further conditions as may be specified in the order are also satisfied

7A (1) The processing—

(a) is either—

(i) the disclosure of sensitive personal data by a person as a member of an anti-fraud organisation or otherwise in accordance with any arrangements made by such an organisation; or

(ii) any other processing by that person or another person of sensitive personal data so disclosed; and

(b) is necessary for the purposes of preventing fraud or a particular kind of fraud.

(2) In this paragraph "an anti-fraud organisation" means any unincorporated association, body corporate or other person which enables or facilitates any sharing of information to prevent fraud or a particular kind of fraud or which has any of these functions as its purpose or one of its purposes.

8 (1) The processing is necessary for medical purposes and is undertaken by—

(a) a health professional, or

(b) a person who in the circumstances owes a duty of confidentiality which is equivalent to that which would arise if that person were a health professional.

(2) In this paragraph "medical purposes" includes the purposes of preventative medicine, medical diagnosis, medical research, the provision of care and treatment and the management of healthcare services.

9 (1) The processing—

(a) is of sensitive personal data consisting of information as to racial or ethnic origin,

(b) is necessary for the purpose of identifying or keeping under review the existence or absence of equality of opportunity or treatment between persons of different racial or ethnic origins, with a view to enabling such equality to be promoted or maintained, and

(c) is carried out with appropriate safeguards for the rights and freedoms of data subjects.

…

10 The personal data are processed in circumstances specified in an order made by the [Secretary of State] for the purposes of this paragraph.

The Eighth Principle (preventing transfer of personal data outside the EEA in the absence of adequate safeguards for data subject's rights) does not apply in the following circumstances:

"SCHEDULE 4
CASES WHERE THE EIGHTH PRINCIPLE DOES NOT APPLY

1 The data subject has given his consent to the transfer.

2 The transfer is necessary—

(a) for the performance of a contract between the data subject and the data controller, or

(b) for the taking of steps at the request of the data subject with a view to his entering into a contract with the data controller.

3 The transfer is necessary—

(a) for the conclusion of a contract between the data controller and a person other than the data subject which—

(i) is entered into at the request of the data subject, or

(ii) is in the interests of the data subject, or

(b) for the performance of such a contract.

4 (1) The transfer is necessary for reasons of substantial public interest.

…

5 The transfer—

(a) is necessary for the purpose of, or in connection with, any legal proceedings (including prospective legal proceedings),

(b) is necessary for the purpose of obtaining legal advice, or

(c) is otherwise necessary for the purposes of establishing, exercising or defending legal rights.

6 The transfer is necessary in order to protect the vital interests of the data subject.

7 The transfer is of part of the personal data on a public register and any conditions subject to which the register is open to inspection are complied with by any person to whom the data are or may be disclosed after the transfer.

8 The transfer is made on terms which are of a kind approved by the Commissioner as ensuring adequate safeguards for the rights and freedoms of data subjects.

9 The transfer has been authorised by the Commissioner as being made in such a manner as to ensure adequate safeguards for the rights and freedoms of data subjects.

Relevant exemptions to the DPA

Section 29 provides various exemptions for the prevention or detection of crime, the first of which is as follows:

"Crime and taxation

(1) Personal data processed for any of the following purposes—

(a) the prevention or detection of crime,

(b) the apprehension or prosecution of offenders, or

(c) the assessment or collection of any tax or duty or of any imposition of a similar nature,

are exempt from the first data protection principle (except to the extent to which it requires compliance with the conditions in Schedules 2 and 3) and section 7 in any case to the extent to which the application of those provisions to the data would be likely to prejudice any of the matters mentioned in this sub-section."

THE INTERNATIONAL CHARACTER OF LAWS OF DATA SECURITY

The law protecting data has, largely because of its borderless nature, been designed at the international level. The current British statute – the Data Protection Act 1998 - is a creature of European law. It springs from the 1995 Data Protection Directive (95/46/EC).

Under the European Communities Act 1972 (due to be repealed by the UK's withdrawal from the EU), directives of the EU are directly applicable in British law, consistently with the Treaties establishing the EU. At the time of writing, EU law still applies, though that will change with withdrawal from the EU.

Beneath the original 1995 EU Data Directive (95/46/EC) are a number of more specific European laws, all of which have influenced British law. Particularly important European laws include:

- The 2002 "e-Privacy" Directive (2002/58/EC), or "ePD". and
- The 2016 General Data Protection Regulation (Regulation 2016/679), or "GDPR", which is due to come into force in May 2018.

In the present context of protection of data arising from the use of vehicles the key features of European law are:

- The fundamental purposes of EU law to protect the right to confidentiality of communications and data, via the Charter of Fundamental Rights of the European Union, given force by the 2009 Lisbon Treaty. These are often referred to as the "third generation" fundamental rights protected by EU law.

- The aim for legal certainty as to the laws applicable to data protection – this is set out in the foundational 1995 Data Directive.

And

- The fundamental requirement for the consent of users of a service to the processing of their data. This is already a guiding principle, enshrined in the British Data Protection Act, for example. But its scope is changing, particularly through the expansion of location data to which smartphone technology and (eg) map and taxi apps have given rise, and which driverless vehicles will push further.

2. DATA LAW IN FLUX

Several phenomena put data protection law in flux. Those include:

- technological progress. As the EU working party on the ePD (July 2016) has recognised, the landscape and complexity of communications technology, as well as the complexity of the commercial interests involved, has magnified. Data crosses borders, as information that would previously have been stored entirely on a device is now diverted to servers, which can be located overseas from the user.

- commercial progress: in particular, the extent to which e-commerce regards information as a key business tool. Much of the tension between users of data on the one hand (in our context, users of autonomous vehicles) and businesses on the other, stems from the desirability to business of obtaining information. Information has profound commercial value, in that it tends to identify future customers. For this reason, many "free" or low-cost services require the customer not to pay financially, in direct terms, but by providing lucrative information. And if the information is not provided by the customer, then the service (or financial discount) will not be provided by the service provider. So commerce dictates behaviour. As the EU's ePD working group put it in July 2016, the "internet of things" also involves the downloading of much data automatically, as a default setting[6], potentially intruding upon the privacy of the user, though without their knowledge or consent[7]. Without adequate justification for that (eg. downloading important security updating software[8]), such intrusion might not be justified.

- political change: in particular, the exit of the UK from the EU, with the removal of direct applicability of EU laws in the UK and the consequent, imminent need to rewrite British data protection laws – either by a complete adoption of the existing EU laws or by redesign.

- legal developments worldwide: including the EU's General Data Protection Regulation, which was prompted by failures of data security in the United States, which had affected EU citizens' data. The law on data is necessarily international. It tends towards protection of individuals.

The General Data Protection Regulation 2016

The European Union's GDPR exemplifies the changing character of data protection law. It grew from large-scale failures to secure European Union citizens' data stored outside the EU.

---

6 EU working group on ePD, July 2016, page 11.

7 EU working party on ePD, July 2016, page 5.

8 EU working party on ePD, July 2016, page 14.

For that reason, the GDPR shifts the focus from the location in which data is held, towards the core issue of whether or not an EU citizen's data security is affected.

The GDPR also sharpens the teeth of enforcement. Data security breaches will be punishable under the GDPR by very large fines.

The GDPR will come into effect in May 2018. The British government has said that the Brexit process will not affect the commencement of the GDPR in the UK (though whether British data law will follow the GDPR path beyond May 2018 is unknown).

In the context of automated vehicles, the GDPR will add several features to British data protection law. Those that might affect autonomous vehicles include the following:

- The protective principles for the processing of personal data are refined[9] to:
    - Lawfulness, fairness and transparency (in relation to the data subject)
    - Limitation to the purposes (of processing the data, which purposes must be specified, explicit and legitimate)
    - Minimisation (of processed data, to achieve the purpose)
    - Accuracy (of data, including the erasure of outdated information)
    - Limitation of storage period (to the time necessary for the purpose)
    - "Integrity and Confidentiality" (ie. secure storage)

---

9 GDPR Article 5(1)

- Data controllers and processors must show how they comply with the all of the data protection principles. This is the overarching "accountability" requirement[10].

- Data protection must be achieved by technical design and by default[11].

- The GDPR prescribes certain lawful purposes for processing, at least one of which must be shown[12]. Those that might apply to a driverless car (assuming the GDPR applied in its present form) are:

    - The consent of the data subject to processing for one or more specific purposes (eg. to identify the destination of an app car, or to convey the owner of a driverless car to their destination)

    OR the need to process data for:

    - Performance of a contract with the data subject (or pre-contractual steps relating to such a contract)

    - Compliance with a legal obligation to which the data controller is subject

    OR perhaps even the need to process data for:

    - Protection of the vital interests of the data subject or another natural person

    - The purposes of the legitimate interests pursued by the controller or by a third party

- The territorial reach of the regulation is wider than the previous law. It applies to data controllers and processors "offering goods

10 GDPR Article 5 (2)

11 GDPR Article 25

12 GDPR Article 6(1)

and services to data subjects who are within the Union"[13], even where the controller or processor is established outside the EU.

- The meaning of "personal data" is wider. For example, information identifying online activity, such as an IP address, can be personal data, as can data that has been "pseudonyomised"[14].

- Consent to data processing must be "given by a clear affirmative act" and "unambiguous". Silence or not altering a pre-ticked box will not suffice[15].

- Information to data subjects should be "easily accessible and easy to understand"[16]

- The data subject has a right to have data provided in intelligible form to him or her and to transmit it to another controller, or where technically feasible to ask a data controller to transmit that data ("data portability")[17]

- Controllers and processors owe enhanced duties to ensure data security, by several measures including (for example) "pseudonymising" personal data as soon as possible[18]

- Where there is a "high risk" in data processing "to the rights and freedoms of natural persons", the controller shall carry out "an assessment of the impact of the envisaged processing operations on

---

13 See preamble to the GDPR at (23) (making the point that "goods and services" implies more than (eg) mere accessibility of a website within the EU) and Article 3.

14 See GDPR preamble at (26) and (30) and Article 32.

15 See GDPR preamble at (32) and Article 7.

16 See GDPR preamble at (58)– including use of "visualisation" to simplify the message – and Article 12.

17 See GDPR preamble at (68) and Article 20

18 See GDPR preamble at (78)

the protection of personal data". That is especially so "where a type of processing" uses "new technologies"[19]

- Member states may "provide for more specific rules to ensure the protection of the rights and freedoms of in respect of the processing of employees' personal data"[20]

The GDPR states its commitment to a balance between the data subject's security and the demands of increasingly sophisticated technologies and international data transfers. But it gives greater weight to the data security of the individual subject[21]. That is its object.

That cannot be an exhaustive list of the possible applications of the GDPR to automated vehicles, as neither that law nor those vehicles exist yet.

It is likely that the GDPR will not be the final word on data security relating to vehicles. If autonomous vehicles evolve as rapidly as many expect, the scale on which personal data is processed will increase significantly.

With that, the demands for effective data security laws will expand. Those laws must be practically enforceable. That will affect not just the law but also the commercial dynamics of the driverless car industry, including insurance.

Information is one of the fields in which the law is likely to continue to develop radically. Driverless vehicles will be a focal area of change.

3. DATA RISKS OF DRIVERLESS VEHICLES

It is not yet possible to predict the variety of ways in which driverless vehicles will affect privacy and, thereby, data protection laws.

19 GDPR Article 35

20 GDPR Article 88

21 See the GDPR preamble, especially at (6) and (7)

But some of the ways in which future, artificially intelligent vehicles (linked to networks and speaking to each other) might raise concerns as to privacy and confidentiality are foreseeable. A few of those are as follows:

The car will store information, of use in relation to accidents, crime etc

The technology of telematics (equipment showing the speed and other circumstances affecting a vehicle in the moments before, during and after a collision) is familiar to those dealing with the aftermath of road traffic accidents.

Data protection law provides exceptions in relation to the detection of crime, and in relation to legally privileged materials.

In the case of driverless vehicles, it is predictable that telematic data will be of greater sophistication, and so of greater use to victims of accidents (as well as to police and to insurers), in the event of collisions or crimes. Balanced against that is the hope that driverless technology might reduce the frequency of both.

But the significance of data will reach much further than telematics information, in the world of driverless cars.

The car will know where you are going

The vehicle will know its destination and – whether directly (in the case of its owner) or indirectly (in the case of a paying passenger) – will be able to identify its passenger.

A person's destination will not, in every situation, be a confidential matter. But it is likely to be automated information which relates to a living individual who can be identified from the information, ie. "personal data" within the meaning of the DPA. So the data protection rules are likely to be triggered. And there are circumstances in which an individual's destination would probably be considered private or confidential information as well.

The EU's Working Party on the ePD has recognised the need to protect location data, even in relation to existing technologies. But it balances against that the ubiquity and utility of location data. The proposed solution is to restrict the use of the data to that required to provide the service that the user has requested (eg. online security software and updates)[22] and to increase data security measures, in particular in relation to "Internet of things" devices (as a driverless car would be)[23]

So, the EU's proposed solution is to recognise social and commercial benefits on the one hand, and the need for security to prevent misuse of information on the other. The risk of such misuse would, however, remain.

The car might want to sell you something

Not the car, perhaps. But the driverless car would probably become (like almost any other networked, consumer device) a means to shop online.

This raises the issue of marketing and, again, the prospect of "data-mining", for the purpose of more efficient selling.

The EU working party's opinion on the use of cookies is instructive on this point. A "cookie" is, in simple terms, a piece of software introduced by a website onto a user's computer, in order to access usage information. The invasion of privacy that would otherwise result is dealt with by the familiar "cookie consent" tick-box that appears on most websites.

---

22 EU working party on ePD, July 2016, page 14.

23 EU working party on ePD, July 2016, page 18.

The growth of machine intelligence, the internet of things and the driverless car are all likely to result in greater "cookie"-like activity.

The EU ePD working party's view is that the service offering a "cookie" should be allowed to make the request for the user's data only if necessary to provide (a) the service asked for by the user (b) security or (b) billing for that service[24].

Cookies are likely to increase in sophistication. They, too, have a positive aspect (allowing the machine to learn and act according to the user's preferences, ie. more efficiently) as well as involving a sacrifice (of control of some personal information). Which leads to another scenario:

Could use of a driverless car be refused if the user withholds access to personal information?

Some personal information (eg. destination) would, practically, need to be given to the car in order for its purpose to be fulfilled.

But the EU ePD working party considered whether, on current technology generally (not in the specific context of driverless cars) that would be true of all personal information. The Working Party's opinion was that use of tracking technology and users "consents" should be used in a restrictive way, in particular not to provide information to third parties, and that consents must be easily comprehensible, so that the information to which the consent relates is clearly identified.[25]

So the degree of protection envisaged by the ePD working party was high.

An implication of this is that the systems of (say) driverless cars must be sufficiently artificially intelligent not only to negotiate the unpredictable hazards of the road, but also to make judgments as to whether inform-

24 EU working party on ePD, July 2016, page 12.

25 EU working party on ePD, July 2016, page 17.

ation is capable – directly or indirectly – of revealing sensitive information, eg. about special categories of data (which would include, under the DPA, information – explicit or implied – as to physical or mental health and as to political opinions).

Which leads to an intriguing point: if the driverless car is required not only to drive but also to make judgements as to the nature and implications of information, then it will be performing functions of a very high intelligence indeed.

What might be the legal implications of that?

SUMMARY

The driverless car will be in command of information that affects not only the physical safety but also the social well-being of its passengers. It will be in a position of unusual power in relation to private information, because it will not only store but be required to reach judgements on the basis of that information – judgements as to privacy as well as physical safety.

At least, that will be so if we wish to retain the level of privacy that the law of data protection currently seeks to achieve, whilst also having the benefit of artificial intelligence.

The driverless car will be a store of potentially very valuable information if, interconnected with other systems, it is used (for example) to access financial records. It is likely to be the target of crime.

The risks of accidental or criminal release of private information, and of financial and reputational loss, will therefore be live. This might be an unfamiliar addition to the list of driving risks, and one of the subject risks in a new generation of motor insurance policies.

The law of data protection will be tested and profoundly reshaped by the arrival of driverless vehicles as might our notions of privacy, and our relationships to each other.

CHAPTER FIVE
ACCIDENTS

INTRODUCTION

Accidents will happen. Driverless cars will arrive on roads still dominated by human drivers. There will still be collisions, at least in the early stages of the technology. Some will be between vehicles, some between vehicles and other road-users (eg. cyclists). Some of those accidents might even result from problems with the new technology.

Not until the more distant future, when road design and vehicle technology have advanced together to accommodate both the new and old technologies – possibly including "zoning" of areas for different categories of vehicles (see the discussion below) – might road accidents diminish to a great extent.

When accidents happen, legal fault (ie. civil liability to compensate the victim) will be determined in a different way to accident claims today.

There will be a different mix of legal relationships: driver, passenger, owner, hirer, employee and bystander will relate to a robotic vehicle in different ways. Legal issues and evidential questions will be different. The law and court procedure will evolve.

This chapter describes the laws of civil liability that are likely to apply to road traffic accidents involving driverless vehicles in that transitional phase, after the introduction of driverless vehicles onto roads still dominated by human-driven vehicles. It deals with those points under the following headings:

- The likely factual background: our new relationships to vehicles
- The present laws that might apply, and
- Current British government thinking.

The objective of this chapter is to familiarise the reader with the near-future, civil liability landscape of driverless vehicles. I call this "the untidy transition" to driverless technology.

1. NEW RELATIONSHIPS TO VEHICLES

Proposals for road transport law reform tend to assume that fully driverless technology is not immediately achievable, and that every vehicle will at all times need a "driver", able to take control if needed.

As discussed in Part One of the book, that "human control principle" is the central assumption upon which international road traffic law (set out in the Vienna Convention) and the British vehicle construction and use regulations are based.

The principle might not survive. The advance of technology (especially in artificial intelligence) suggests that a robotic driver might one day be safer than a human driver.

If that is the likelier prospect, then the "intervening human driver" (as I shall call him) might turn out to be more of a nuisance than a help. He might take over control when it would be safer not to do so. He might overreact and make injury more likely.

But technology has yet to advance to the stage where it is demonstrably safer than a human driver. So we must deal with the facts as they are.

In the untidy transition, technology will have yet to convince the public of the superior safety of robotic road vehicles.

As a result, those vehicles will not be fully driverless but "semi-autonomous" vehicles ie. always capable of control by human intervention.

In factual terms, our relationships to semi-autonomous vehicles (and consequently to others) are likely to be as follows.

Driver

As the law stands, the driver of a conventional vehicle is required to take reasonable care for anyone whose safety might reasonably foreseeably be affected by the driving of the car (the "reasonable foreseeability" test, ie. the standard of care in the law of negligence).

That presumes uninterrupted human control of the vehicle. There has been discussion of imposing a more demanding standard of care upon a driver of a semi-autonomous vehicle, to reflect the increased hazard that only partial control of a vehicle would bring[1]. But that suggestion has not yet been adopted, and might not be adopted.

Whether or not the civil standard of care upon a driver is altered, the "driver" would remain the ultimate controller of a semi-autonomous vehicle. He would owe a duty of care (to whichever standard) to anyone – passenger, owner, bystander – who suffered injury or loss (eg. repair costs) as a result of misoperation of the vehicle.

The driver would also look to others to guard his or her safety. Those might include:

- If not him, the person responsible for the safe upkeep of the vehicle (particularly if the driver uses the car for work for another, eg. a taxi company)
- The manufacturer of the vehicle and of its systems (including its software)
- The driver(s) of other vehicles on the road (whether new or old technology vehicles)

1 In the Dept of Transport's first "Pathway to Driverless Cars" paper, in February 2015. That paper took a more prescriptive approach to preparing the statutory and regulatory path than the subsequent government papers (July 2016 and January 2017), which recommend several "waves of reform" depending upon the performance of the technology.

- Other road users who might affect the safe operation of the vehicle (eg. pedestrian stepping suddenly out into its path) and

- The body responsible for the safe upkeep of the road.

(I shall deal with the last category – roads – in another chapter)

Passenger

In the context of fully "driverless" vehicles, the distinction between "driver" and "passenger" is likely to disappear: all occupants of a fully driverless car would be "passengers" and there would be no human "driver".

But that is a far-future possibility (discussed in Part Three of the book). The situation with a semi-autonomous vehicle is different: a human driver is always required, to take control whenever needed.

So a passenger in a semi-autonomous vehicle would probably look to the following people to guard his or her safety:

- The driver of that vehicle (and anyone vicariously liable for the driver, eg. his employer)

And (just as in the case of the driver, above):

- The person responsible for the safe upkeep of the vehicle (probably its owner)

- The manufacturer of the vehicle and of its systems (including its software)

- The driver(s) of other vehicles on the road (whether new or old technology vehicles)

- Other road users who might affect the safe operation of the vehicle and
- The body responsible for the safe upkeep of the road.

Owner

A real possibility is that the number of vehicles in private, individual ownership will decrease. This would not be the result solely of driverless technology.

Increasingly, in some areas, the availability of alternatives to owning a car (calling a taxi from a smartphone app, hiring a car from a company providing hire cars parked nearby) is already calling the economy of individual car ownership (which includes the costs of purchase, financing, maintenance, fuel, road tax, parking charges and the high level of vehicle value depreciation) into question.

The economic direction might therefore be away from individual car ownership – at least in areas well-served by alternative transport, chiefly public transport. So the initial growth areas for autonomous transport are likely to be cities.

In the transitional phase of semi-autonomous vehicles, there might still be a great many, individually-owned vehicles on the road. Those would probably, initially, be mainly non-autonomous vehicles.

But as energy technology progresses (and – should this direction in environmental policy continue – fossil fuel-driven vehicles are replaced by electrical alternatives), it seems likely that autonomous vehicles will increase in number. Autonomous devices are a feature of some types of electrical vehicle.

In the transitional phase, therefore, it seems likely that:

- The national fleet (as it were) of road vehicles would consist to a greater extent of semi-autonomous, electrically-powered vehicles

- Were that trend to increase, people might move away from individual ownership of those vehicles, for reasons of economy and convenience, and prefer to use alternatives and to hire cars from "app" services, as the need arises. If so, that might be influenced by the easy availability of such vehicles, so might be a trend focussed initially in cities.

- The trend of private ownership might survive – but probably only if the new generation of cars are no more expensive to own than traditional, fuel-driven cars.

Hirer

The hirer of a vehicle would usually be the driver. In the case of a semi-autonomous vehicle, the hirer would then look to the following people to guard his or her safety:

- The person responsible for the safe upkeep of the vehicle (the hire company)

- The manufacturer of the vehicle and of its systems (including its software)

- The driver(s) of other vehicles on the road (whether new or old technology vehicles)

- Other road users who might affect the safe operation of the vehicle (eg. pedestrian stepping suddenly out into its path) and

- The body responsible for the safe upkeep of the road.

Employee

In the transitional phase, vehicles would become increasingly automated, but the requirement for a "driver", always able to take control, would remain.

An employee required to use a semi-autonomous vehicle as work equipment would therefore be required to "drive" it. But he or she might also be required to do other work while the vehicle was in motion, eg. performing administrative tasks such as checking delivery lists.

So, in the event of an accident, the questions of when an employee tasked with driving a semi-autonomous vehicle ought reasonably to have taken control and, if so, whether such intervention would on the balance of probabilities have prevented or reduced injury and damage, will be live.

Employees would look to the following people to guard their safety:

- Their employer: in particular to provide:
    - A safe system of work – including sufficient training and a safe system for the operation of autonomous work vehicles (including the conditions of employee supervision of and intervention in automated driving).
    - Safe work equipment – including safe autonomous vehicles and responsibility for any defects in those vehicles causing personal injury, under the Employers' Liability (Defective Equipment) Act 1969.
    - Compulsory insurance "against liability for bodily injury or disease sustained by his employees, and arising out of and in the course of their employment in Great Britain in that business" under the Employers' Liability (Compulsory Insurance) Act 1969.

The employer's work equipment duties might avoid any need for an employee to sue a manufacturer in relation to defects in a vehicle.

And a recent judgment suggests (albeit only at first instance and in a different legal context) that the traditional treatment of drivers used by some passenger transport businesses (in that instance private hire taxi drivers) as self-employed might shift, to provide them with some employees' protections[2].

Absent such expansion of "employment" by the higher courts, however, self-employed drivers would look for their protection to the same people as drivers (see above).

Employed drivers would also look to the following people to take care of their safety:

- The driver(s) of other vehicles on the road (whether new or old technology vehicles)
- Other road users who might affect the safe operation of the vehicle (eg. a pedestrian stepping suddenly out into its path) and
- The body responsible for the safe upkeep of the road.

Bystander

A bystander would look to the following people for his or her protection:

- The driver(s) of other vehicles on the road (and their employers)
- The body responsible for the safe upkeep of the road.

---

2 *Aslam & another v Uber B.V. & others,* 28 October 2016, London Central Employment Tribunal

The bystander would also look – indirectly – for protection to the motor insurer of vehicles, via the compulsory road traffic insurance regime for third party injury and property damage.

2. PRESENT LAWS THAT MIGHT APPLY TO ACCIDENTS INVOLVING SEMI-AUTONOMOUS VEHICLES

The laws applying to the various legal relationships between human and semi-autonomous car, in a claim arising from an accident, are likely to include the following.

Driver

Negligence and the Current Standard of "Reasonable Care"

A driver of a vehicle owes a duty to drive with reasonable care for all those whose safety might reasonably foreseeably be affected by his driving.

The standard of care is that of the ordinarily skilful driver. But it is a high duty, "to reflect the fact that a car is potentially a dangerous weapon"[3].

On current law, the driver is required to be have control of the vehicle at all times. A current vehicle that allows a function to become automated (say, parking assist or cruise control) tends not to disturb that control, as it operates either while the driver is still directing the path of the car by steering, so is necessarily concentrating upon the road (cruise control) or occurs in a short space of time, at low speed and subject to an automated stop should an obstacle be detected (park assist).

3 *Lunt v Khelifa* [2002] EWCA Civ 801, Latham LJ at para.20

The Liability Gap when "driving" semi-autonomous vehicles

The government's January 2017 response to the consultation in relation to driverless cars proposes the extension of compulsory insurance to semi-autonomous cars being "driven" in automated mode, and the retention of the EU right to issue a claim directly against a motor insurer[4].

In the government's proposal:

> *"When a crash is determined to have been caused by an [autonomous vehicle], where the [automated driving feature] was active, the insurer would be liable to pay compensation to the innocent third party victim. They would also pay out to the motorist if injured in the vehicle if the ADF were active … Where the manufacturer is found to be liable, the insurer will be able to recover against the manufacturer under existing common law and product liability laws"*[5]

So the motor insurer would be responsible for satisfying a judgment obtained by an injured first party driver and an injured third party, but would have the right of indemnity against the manufacturer of the car if the ADF was at fault.

So, under the January 2017 government proposal, the insurer will pay for third party injury caused either by a liable human driver (as it does now) or by the automated vehicle.

That goes some way to providing a practical solution to a concern about effective compensation, namely that a person injured in an accident attributable to a semi-autonomous vehicle moving in automated mode

---

4 Dept of Transport's "Pathway to Driverless Cars", government response to consultation, January 2017. The insurance proposal also identifies limited exceptions to the insurer's cover, namely that the crash was caused by (i) the driver making unauthorised alterations to the car's operating system or (ii) failing to update the car's software. Hacking by a third party would not exclude cover. See page 11, 3.13.

5 Dept of Transport January 2017 paper, p.11 at para's 3.13 and 3.14

might find that he could prove neither driver negligence nor machine defect, so fail to recover any compensation at all.

But, in effect, that passes both difficulties (of payment to the victim and of seeking contribution) to the insurer. And it is a solution of payment; it does not answer the underlying causation questions of fact and law, which an insurer must address in order to assess the usefulness of contribution proceedings against the manufacturer. It leaves those problems to the insurers and to the courts.

In particular, it does not address the grey area of "taking control". When should a human driver reasonably take back control of a semi-automated vehicle, running in automated driving mode? This is a feature unknown to – and an area untested in – British road transport law.

The Highway Code

Section 38(9) of the Road Traffic Act 1988 describes the Highway Code as "the code comprising directions for the guidance of persons using roads".

Its effect in law is set out in Section 38(7) of the 1988 Act:

> *"A failure on the part of a person to observe a provision of the Highway Code shall not of itself render that person liable to criminal proceedings of any kind but any such failure may in any proceedings (whether civil or criminal, and including proceedings for an offence under the Traffic Acts, the Public Passenger Vehicles Act 1981 or sections 18 to 23 of the Transport Act 1985) be relied upon by any party to the proceedings as tending to establish or negative any liability which is in question in those proceedings."*

The current version of the Highway Code explicitly advises against the use of driver assistance systems as a substitute for driver concentration, at rule 150:

> *"Do not rely on driver assistance systems such as cruise control or lane departure warnings. They are available to assist but you should not reduce your concentration levels."*

That rule cites Regulation 104 of the Road Vehicles (Construction and Use) Regulations 1986 ("Driver Control"), quoted in Chapter 1.

Passenger

An individual or business carrying passengers in a vehicle owes a duty of care to take reasonable care for the safety of that passenger (whether they are carrying the passenger gratuitously or for reward). So they have a duty to ensure that the vehicle is driven with reasonable care.

Someone responsible for the safe state of a vehicle (whether owner or – for example – transport company) owes a duty to take care for the safety of the vehicle. The duty in tort, to the public at large, is to take reasonable care. However a carrier of passengers owes a "high degree of care", which has been described as "the duty of exercising all vigilance to see that whatever is required for the safe conveyance of their passengers is in fit and proper order"[6].

Consumer Rights

In addition to any express contractual rights, a paying passenger would be entitled to statutory consumer rights. Under the Consumer Rights Act 2015 (discussed below), in a services contract, services are to be per-

---

6 *Readhead v Midland Railway* (1869) L.R. 4 Q.B. 379, 393, The analogies with tram, stagecoach and railway cases are likely to multiply as semi-autonomous vehicle accident cases come before the courts.

formed with reasonable care and skill – Section 49. Damages are amongst the remedies available for breach (Section 54(7)).

Unlike the owner's rights relating to the quality of goods and digital content (discussed below), "safety" is not an aspect of services to be taken into account explicitly under the Act. But, in the context of a taxi journey in an autonomous taxi (attended by a driver), courts might consider the passenger's "safety" as implicit in the "reasonable care" of the taxi service. The courts would also take account of the high standard of care owed by carriers to passengers in relation to the safety of the vehicle (above).

Other rights of passenger

A passenger also has rights of action for damages under:

- The common law of negligence: in relation to risks of injury and damage reasonably foreseeable by the carrier. These include warnings of particular dangers. As discussed in the "data" chapters, this might include the risk of breaches of duties relating to confidential and personal information.
- The Occupiers Liability Act 1957: in relation to such risks arising from the state of the vehicle (eg. tripping injuries).

Owner

Consumer Rights

If the owner of a vehicle is -

> *"an individual acting for purposes that are wholly or mainly outside that individual's trade, business [which includes the activities of any government department or local or public authority], craft or profession"*

– then that person is a "consumer" whose contractual rights described below have, since 1 October 2015, been provided by the Consumer Rights Act 2015.

The person who owes duties to the consumer is the "trader", who is defined as:

> *"a person acting for purposes relating to that person's trade, business, craft or profession, whether acting personally or through another person acting in the trader's name or on the trader's behalf"*

Part 1 of the CRA 2015 applies to "consumer contracts, digital content and services". "Goods" and "digital content" are distinct: "Goods" means "any tangible moveable items" while "digital content" means "data which are produced and supplied in digital form".

The statutory rights of likely application to the rights of an injured, non-business buyer of an autonomous car under the CRA 2015 are as follows:

- Both the goods and digital content are to be of satisfactory quality (including safety – discussed further below) – Sections 9 and 34

- If "the goods are an item that includes digital content" and "the digital content does not conform to the contract to supply that content", the goods will not conform to contract – Section 16

A breach of either section can be remedied in ways including the consumer seeking damages (see Sections 19(11) in relation to goods and 42(7) to data). That would be the familiar remedy for personal injury of the owner – or the basis of a contribution claim in the event that the owner is sued after an accident, for example by an injured passenger.

How would the courts fix the appropriate standard of "safety", when applied to a partially-autonomous car, under human control?

Such a vehicle might be marketed on the basis that it can be left – subject to the driver's overarching control – relatively unattended while driving. That does not appear to be the capacity of the technology, yet. Such a description might risk contravening the Highway Code and Road Vehicle (Construction and Use) Regulations, as they stand in relation to driver control.

And, against that background, a trader might create difficulties for itself by describing a vehicle in language giving the impression that it is, at any time, capable of safe, independent driving.

Such statements would be circumstances relevant to the quality of the goods[7]. So there might be particular tension around terms like "driverless", in the context of cars actually required – both technologically and by regulation to be under human control.

An accident involving a vehicle described in those terms might result in a damages action based in part upon the CRA 2015.

Product Liability and other consumer protections

Beyond contractual rights, the owner is also owed duties:

- in negligence (the duty of anyone delivering a product in its final form to the consumer – designer, manufacturer, retailer, safety testers and others to take reasonable care to protect the consumer from injury and property damage that they can reasonably foresee might result from use of the product: the duty of care in *Donoghue v Stevenson*[8])

- in the statutory law of product liability. This provides another line to a remedy where a product is defective but liability in negligence

7 See CRA 2015 subsections 9(5) and (6) and 34(5) and (6) – whereby the circumstances relevant to quality of goods and of digital content include "specific characteristics of the goods made by the trader, the producer or any representative of the trader or the producer" and "in particular, any public statement made in advertising or labelling".

8 [1932] A.C. 562, 599, Lord Atkin.

would be difficult to prove (eg. negligence lies with an inaccessible defendant, such as a manufacturer far overseas). The Consumer Protection Act 1987 (Part 1) is the English statute, though it is a creature of European Union law. Again, the protection is for consumers – not for businesses. Section 2 provides strict liability where "any damage is caused wholly or partly by a defect in a product[9]", subject to certain defences. Absent such defence, liability is placed strictly upon the (a) producer (b) anyone who has held himself out as producer, by putting his name, trade mark or other distinguishing mark on the product and (c) an importer from outside the member states of the European Economic Area.

Section 3(1) of the CPA 1987 defines "defect" as follows:

3 Meaning of "defect"

(1) Subject to the following provisions of this section, there is a defect in a product for the purposes of this Part if the safety of the product is not such as persons generally are entitled to expect; and for those purposes "safety", in relation to a product, shall include safety with respect to products comprised in that product and safety in the context of risks of damage to property, as well as in the context of risks of death or personal injury.

(2) In determining for the purposes of subsection (1) above what persons generally are entitled to expect in relation to a product all the circumstances shall be taken into account, including—

(a) the manner in which, and purposes for which, the product has been marketed, its get-up, the use of any mark in relation to the product and any instructions for, or warnings with respect to, doing or refraining from doing anything with or in relation to the product;

9 Section 1(2) CPA 1987: ""product" means any goods or electricity and … includes a product which is comprised in another product, whether by virtue of being a component part or raw material or otherwise"

(b) what might reasonably be expected to be done with or in relation to the product; and

(c) the time when the product was supplied by its producer to another;

and nothing in this section shall require a defect to be inferred from the fact alone that the safety of a product which is supplied after that time is greater than the safety of the product in question.

The defences are set out in Section 4:

4 Defences

(1) In any civil proceedings by virtue of this Part against any person ("the person proceeded against") in respect of a defect in a product it shall be a defence for him to show—

(a) that the defect is attributable to compliance with any requirement imposed by or under any enactment or with any [EU] obligation; or

(b) that the person proceeded against did not at any time supply the product to another; or

(c) that the following conditions are satisfied, that is to say—

(i) that the only supply of the product to another by the person proceeded against was otherwise than in the course of a business of that person's; and

(ii) that section 2(2) above does not apply to that person or applies to him by virtue only of things done otherwise than with a view to profit; or

(d) that the defect did not exist in the product at the relevant time; or

(e) that the state of scientific and technical knowledge at the relevant time was not such that a producer of products of the same description as the product in question might be expected to have discovered the defect if it had existed in his products while they were under his control; or

(f) that the defect—

(i) constituted a defect in a product ("the subsequent product") in which the product in question had been comprised; and

(ii) was wholly attributable to the design of the subsequent product or to compliance by the producer of the product in question with instructions given by the producer of the subsequent product.

There is legal debate internationally as to whether the generic development of "driverless" technologies might allow the argument that the defect is not actionable, because it results from necessary experiment for the benefit of society.

Whether that argument is open on the English statute is questionable. Section 4(1)(e) of the CPA 1987 deals with a defect that scientific and technical knowledge could not have foreseen, ie. the undiscovered glitch.

There is a difficulty with that argument in the present context ~~is clear~~: the glitch (incomplete automated control) helps to define the aim of the technology, namely the production of a driverless car. So the problem is not only foreseeable; it is immediately apparent. So a producer, supplier or importer might face challenges to the defence of a semi-autonomous car accident claim under the CPA 1987 on the ground that that they could not have foreseen incomplete robotic control, or inadequate human control.

Whether components of driverless technology (eg. machine learning) will produce as yet unknown defects is another question. The technology is bound to surprise in some ways, so unwelcome surprises are a possibility.

Were that issue to arise, factors such as the adequacy of warnings as to the limits of the technology (probably including the language of advertising) would also be relevant.

Contractual Rights for non-consumers

A business purchasing autonomous vehicles for use in its business would not have rights as a consumer (nor would a government department). Its rights would be contractual, as between it and the seller of the vehicle.

The business would have to comply with current regulation – particularly regulations relating to work equipment, to vehicles and to employee safety. Those regulations would place onerous duties upon it.

As discussed below, the duties owed to employees present a particular challenge to semi-autonomous vehicles, as they highlight the dangers of a driver "multi-tasking" while the vehicle is in motion.

On the current law and the guidance of the Highway Code, multi-tasking is prohibited. As this book is going to press, the government is proposing stricter manufacturing standards and criminal measures aimed at preventing the use of mobile phones by drivers, on the ground that such use diminishes driver concentration and threatens road safety.

So the tendency of current law is against "driver detachment" through technology (in my phrase). "Semi-autonomous" or "assisted" driving are the phrases on the positive side of the argument. The technology has not yet reached the level of reliability to overcome the safety concerns. The human parent cannot yet hand over control to the infant technology. The transitional phase will be the difficult growing period.

Hirer

The consumer hirer (ie. not hiring for business use) has protections under the Consumer Rights Act 2015.

The goods provisions of the CRA 2015 apply to contracts for the hire of goods (see section 3). The supply of digital content and services would also appear to be covered (sections 33 and 48).

A hirer of a semi-autonomous vehicle would, in addition to his or her contractual rights, also be owed the duties in the common law of negligence.

Employee

An employer owes duties of care to an employee:

- At common law, including the duties to provide safe work equipment and to provide a safe system of work. The standard of care is reasonable care, though in the employment context that is to be measured carefully by the factual circumstances (for example, statutory guidance might have been observed but negligence still result). The employer cannot delegate the common law duties to anyone else.
- Under statute, through the Employers Liability (Defective Equipment) Act 1969.

The 1969 Act provides (at Section 1) that

> *"(1) Where ...*
>
> *a. an employee suffers personal injury in the course of his employment in consequence of a defect in equipment provided by his employer for the purposes of the employer's business; and*

> *b. the defect is attributable wholly or partly to the fault of a third party (whether identified or not),*
>
> *the injury shall be deemed to be also attributable to negligence on the part of the employer (whether or not he is liable in respect of the injury apart from this subsection), but without prejudice to the law relating to contributory negligence and to any remedy by way of contribution or in contract or otherwise which is available to the employer in respect of the injury."*

"Defect" has been construed widely: "*to include everything which renders the equipment … unfit for the use for which it was intended when used in a reasonable way and with reasonable care*"[10].

An employer cannot contract out of the 1969 Act (Section 1(2)).

Until October 2013, the employer also owed duties under health and safety regulations, which could be relied upon both under the criminal law and in civil actions for damages for injury. Those regulations included the Provision and Use of Work Equipment Regulations ("PUWER").

However, the Enterprise and Regulatory Reform Act 2013 removed civil liability for breach of almost all applicable health and safety regulations[11], so current health and safety regulations are unlikely to be relevant to a claim for damages arising from the use of an autonomous vehicle.

On current law, the driver is required to be able to take control at any time. So a job that requires (whether expressly or by implication, eg. by time constraints on tasks) the employee to carry out another task whilst driving (eg. checking a list of deliveries) might put the employer in breach of its civil legal obligations to the employee – and potentially in breach of public (ie. criminal) health and safety regulations as well.

10 *Yarmouth v France* (1887) 19 Q.B.D. 647, 658, Lindley L.J.

11 The exceptions being regulations relating to compulsory maternity leave, under the Health and Safety at Work etc. Act 1974 (Civil Liability) (Exceptions) Regulations 2013, SI 2013/1667.

The employment example might, again, highlight a practical difficulty of "semi-autonomous" cars: either the vehicle is reliably safe (which would imply no need for human intervention, ie. full autonomy) or it is, in practical terms, a vehicle dependent upon human control. So long as a driver is required to intervene when needed, it might be difficult to argue that a multi-tasking driver (especially an employee) is reasonably safe.

Bystander

A bystander to the vehicle (eg. pedestrian, cyclist) will be owed:

- duties in negligence by the driver, as well as by the manufacturer, supplier etc, in relation to any reasonably foreseeable risks of injury to him or her
- duties by the body charged with the upkeep of roads (see the "Roads" chapters)
- indirect duties such as the provision of compulsory insurance cover, via the Road Traffic Act.

Warnings and Exclusions

Producers etc are prohibited from contracting out of their obligations that might arise from accidents, under the various Acts (eg. by section 7 of the CPA 1987 and by various sections of the CRA 2015 – chiefly Section 65 which prohibits a trader "*by a term of a consumer contract or by a consumer notice exclud[ing] or restrict[ing] liability for death or personal injury resulting from negligence*").

Those provisions might limit the extent to which (for example) manufacturers might describe a semi-autonomous vehicle in terms suggesting that it has "driverless" capabilities, and yet seek to limit their liability by contractual terms suggesting consent to the risks of distracted driving.

A similar provision applies under section 29 of the Public Passenger Vehicles Act 1981, prohibiting the limitation or exclusion of liability for bodily injury in a contract for the conveyance of a passenger in a public service vehicle.

3. CURRENT BRITISH GOVERNMENT THINKING

The British government's thinking on civil liability for accidents involving semi-autonomous vehicles (the current state of the technology) has altered over time:

- Government papers in February 2015 and July 2016 (the latter after a change of government in the May 2016 general election) indicated two slightly differing approaches. The February 2015 paper favoured a prescription by the government of a detailed, new set of laws applying to driverless vehicles – including the possibility of a higher standard of care to apply to the "driver" of an autonomous (ie. partially driverless) vehicle than to the driver of a human-controlled vehicle. The July 2016 paper – launching a consultation – maintained that possibility but adopted a less prescriptive approach, envisaging an essentially flexible and light-touch approach, whereby the government would legislate in "waves", in response to technological developments. Seeking to build the domestic driverless car manufacturing business appeared to be the key factor in government thinking.

- The January 2017 paper (described above) maintained the flexible, "waves of reform" approach, though with the reassurance of the expansion of compulsory insurance cover to accidents occurring in automated mode.

The January 2017 paper also indicated that a feature of European Union law would probably survive Brexit – namely the right directly to sue a motor insurer (currently British law under the European Communities (Rights against Insurers) Regulations 2002).

That is similar to the British government's position in relation to the May 2018 commencement of the EU General Data Protection Regulation 2016. As in data law, much British motor insurance law is rooted in European Union law. While that might well change in the years after Brexit, for the moment at least, the law is staying on that path.

SUMMARY

The mixture of causes of action described above create a complex picture.

Current accidents, between entirely human-driven vehicles, turn almost exclusively upon the law of negligence. Vehicle defects are seldom relevant.

In the world of semi-autonomous vehicles, however, the mechanisms and design of the vehicles, as well as the language of the terms on which they are advertised, bought and hired will all come into play.

There will be a proliferation of potentially relevant parties. Actions for contribution and indemnity, invoking a variety of legal relationships and laws, will result.

This will be a complex, untidy but fascinating interim phase.

CHAPTER SIX
CRIME

INTRODUCTION

This chapter considers three aspects of criminal law that might be affected, as driverless technology is first introduced to public roads in the UK.

Still in the "untidy transition" period, this will involve autonomous cars that are not yet fully driverless, as well as a still-dominant fleet of entirely human-controlled vehicles.

Data security has already been discussed. It is impossible to predict how road traffic crime might be altered by autonomous vehicle technology. But certain areas spring to mind. This chapter examines three such areas.

The three aspects of criminal road traffic law examined in this chapter are the laws relating to –

- Regulation of speed

- Vehicle safety regulations (including regulations as to vehicle construction and employee safety)

  and

- Road behaviour (or how robots and human road-users might mix criminally).

REGULATION OF SPEED

The speed of motor vehicles is, arguably, the most prominent area of regulation of driver behaviour.

Excessive vehicle speed is a major contributor to road casualties. Efforts to curtail speed (eg. by speed cameras) provoke strong reactions from some drivers.

Current driving technology could restrict maximum speed. A speed limitation device does not require sophisticated artificial intelligence, or technology beyond current levels. A car with cruise control probably also has a speed limitation device. Some vehicles (vehicles with more than 8 passenger seats and goods vehicles with a maximum laden weight of more than 3.5 tonnes) must be fitted with a speed limiter.

Distance-governing devices (preventing a car from travelling too close to the vehicle in front) are less common but are currently available.

In theory, then, a government could seek to legislate to prevent cars travelling beyond a maximum speed. Cars could, in theory, be limited mechanically to a top speed of 70 mph.

That would prevent some practical difficulties, however. The imposition of a maximum speed would not prevent the car being driven illegally fast on roads with a lesser speed limit (eg. in a 30 mph residential zone), nor at speeds within the limit but still dangerously fast for the circumstances (eg. wet road, a sharp bend).

So the issue of speed limitation is dependent upon technology, some of which we still lack.

However, it is possible that, in the near-future, there might be zones (eg. in some parts of cities) where the only vehicles allowed to operate are driverless. And those cars might be networked to communicate with each other and to travel only at safe distances from each other. That might happen during the "transition" period.

If that were so, the likelihood is that those driverless cars would also be speed-limited to a single maximum speed, appropriate limit to that particular zone – say 20 mph in an urban zone also used by pedestrians, cyclists etc.

And, if networked, those vehicles might also be limited not just to the maximum speed for that zone, but to the maximum safe speed for any particular moment in time.

For example, a car aware of the presence of nearby pedestrians with pets (either from its own sensors or from information transmitted by other nearby cars) might be able to adjust its speed downwards accordingly, so as to be prepared in the event of (say) a dog suddenly running in front of it.

The limitation of speed would also have implications for the inhabitants of the vehicle. Deceleration from an already low speed, even in an emergency halt, would in all likelihood reduce the possibility of injury to the occupants of the car.

A car under human control must be designed and constructed so as to give the driver a full view of the road and traffic ahead of the vehicle[1]. But seats within a fully-driverless car need not face the direction of motion, so more ergonomic solutions to the familiar problem of whiplash injury might be available.

The "zoning" of traffic within even small areas of towns might become a prominent feature of the driverless age. Just as the segregation of cyclists and motor traffic aims to reduce conflict between the two, so might road use in the future be planned with greater separation of users in mind.

The "emergency stop" might, in such an environment, become a much less likely phenomenon. A driverless vehicle fully aware of its surroundings, travelling at speeds constantly adjusted to those surroundings and applying margins tending towards safety, might be more likely to avoid collision by slowing or moving. The sudden, jarring halt might prove to be a relic of a past age.

There is reason to believe that those features (the launch of fully driverless "pod" type of cars, the zoning of small areas according to types of road-user and the mechanical restriction of speed by a combin-

1 Road Vehicle (Construction and Use) Regulations 1986, regulation 30 – "Vision".

ation of hardware and software) might start to appear in the transitional phase. The "zoning" aspect tends to suggest that this is likely to happen in urban areas.

But, if that new mode of transport were to evolve successfully, there seems no reason why similar principles could not also be used on different roads, spanning longer distances. Why could motorway lanes not also be "zoned", for use by different categories of vehicle (eg. goods haulage and public transport in the inside lane, pods in the lane or lanes to the right of that)?

If speed were controlled mechanically, accidents and consequent disruptions to that road system might be far less frequent.

And, eventually, the need to regulate speed by criminal laws applicable to drivers might disappear. A car capable of speeding could be an obsolete and dangerous machine.

VEHICLE SAFETY REGULATIONS

As discussed in the "Accidents" chapter, breaches of health and safety regulations can no longer be relied upon to establish entitlement to damages in a civil claim.

But the regulations do remain in effect for the purpose of ensuring public safety, particularly the safety of employees.

And for the reasons explored in the "Accidents" chapter, the safety of employees working in vehicles highlights some of the legal challenges of semi-autonomous cars.

The central unknown factor of driverless cars is whether or not technology will deliver a sufficiently reliable, artificially intelligent vehicle to deserve the label "driverless", and to command the confidence of its passengers and those around it.

But that seems increasingly to be a likelihood.

Current vehicle safety regulations require the control of a human driver. By analogy, regulation of hazardous machinery used at work tends to require an "override" switch. For example, regulation 15 of the Provision and Use of Work Equipment Regulations 1998 ("Stop Controls") requires every employer to ensure

> *"that, where appropriate, work equipment is provided with one or more readily accessible controls the operation of which will bring the work equipment to a safe condition in a safe manner."*

There is, in theory, no reason why driverless technology cannot feature an "override" switch. The question might be as to who should operate it.

Should, for example, the decision to override a vehicle be with its occupants? Or should it be with an overseeing authority?

To give an example – a passenger on a tube train, perceiving a danger, presses an alarm button. Does that stop the train? Not usually – it alerts the driver, who would usually stop the tube train at the next stop.

If driverless cars were to be "zoned" and provided within that zone by a company, then might they be supervised by an operations room, and the "override" switch not stop a car instantly, but communicate the perception of danger to the operations staff, who would then decide how to act?

Or could an override switch operated on one vehicle act to stop not just that but all nearby vehicles? Would that solution be vulnerable to crime (eg. a thief deliberately stopping one vehicle in order to isolate potential victims in others)? Might that favour the "control room" approach?

Might the protection of driverless vehicles from some criminal behaviour involve an infrastructure of control rooms more akin to current public transport than to the individually-owned cars of today?

Might driverless cars necessarily involve individuals passing more information to central authorities and companies, such as that? And how would data law be rewritten to provide sufficient protections for individual privacy?

The infancy of driverless cars is likely to be one of greater physical safety in the sense of collision injuries, but also a period of discovering new vulnerabilities. Society's response will be a reshuffling of existing laws as well as the invention of others.

DRIVERLESS CRIME

Hacking of the car's systems has been considered in other chapters.

There has been speculation as to the vulnerability of driverless cars to "road rage" type of attacks, eg. from other drivers frustrated by their speeds.

It has been suggested that driverless cars might also be vulnerable to road traffic accident fraud, by criminals attempting to lure them into collisions in order falsely to claim compensation for injuries that they did not suffer.

The "road rage" type of attack might be a prospect. But it is a risk that can be avoided or mitigated – probably not by high-technology means. "Zoning" of vehicles – at least in the initial stages of driverless vehicles – might prevent friction between the manner in which the new vehicles and some older cars are driven. Segregation of road users into safe areas has had some success in relation to vehicles and bicycles in some urban areas.

The road traffic fraud risk can also be reduced – probably dramatically – by technological advances. Currently, cameras GPS and "telematic" devices are used in some motor vehicles (eg. police cars) to record details of that vehicle's speed and location.

Similar devices are also in use in some private cars, at the suggestion of motor insurers, as well as in the vehicles of many professional drivers, to record images and details in the event of an accident. Internal and external cameras are in use on public transport.

The likelihood is that autonomous and – later – driverless cars will be much better equipped in that respect than their current equivalents. An autonomous vehicle will be able to record the circumstances of an incident, in detail, through several different media. It will probably also be able to notify the authorities (ambulance, police etc) very quickly, in the event of an incident requiring help.

It will probably be able to acquire that information – images, location data, audio recording – simultaneously with the incident. The car will be an effective multi-tasker, in what for a human would be a highly distracted moment. So the car is likely to be a much more effective gatherer of information to combat crime.

This brings us back to the subject of data.

In order to provide security, it is likely that the car of the future will need to process and store data. That has not been a feature of vehicles. That will change our sense of privacy – and the laws relating to privacy. How society decides to tackle those issues will be at the centre of the new law of driverless cars.

SUMMARY

Driverless cars carry some risks. But a self-driving car (and its precursor, the semi-autonomous car) will enable us to remove the dangers of driver distraction and excessive speed, which now can only be tackled by the criminal law.

Its capacity to detect and record features of an incident instantaneously should deter much future vehicle crime.

Automating speed will leave our roads and vehicles safer, and take the control of vehicles largely out of the control of the criminal law.

CHAPTER SEVEN
ROADS

INTRODUCTION

Driverless cars are not a recent idea. In the 1950's, American inventors planned a highway in which cars were directed by electric devices hidden within the road. Despite experimentation, the scale of work required was too great for the idea to be viable.

Roads do not play the main role in many discussions of driverless technology[1]. The important technology today is perceived mainly to be in the car itself (its hardware and software).

But roads have an important part to play. In particular:

- Road organisation and, as discussed already, segregation of different types of road traffic (pedestrians, cyclists, self propelled vehicles – automated and not) into zones probably has an important role to play in safety, and

- Connectivity of automated vehicles to networks will be a prerequisite for driverless technology. Driverless cars will need to be able to talk to each other. That will require reliable networks – particularly close to roads. So, in effect, we will need reliable, nationwide network coverage.

This chapter discusses the current law relating to the "road" part of "road traffic", and how it might adapt to driverless technology.

---

1 One of the admirable exceptions being Chapter 7 ("First there were electronic highways") of "Driverless" Intelligent Cars and the Road Ahead" (MIT Press, 2016), especially page 112 onwards – "The Golden Age". The authors, Hod Lipson and Melba Kurman, point out that the notion of driverless cars goes back even further, to the World's Fair in 1939.

The Highways Act 1980

The key statute is the Highways Act 1980.

The parts of the Highways Act most relevant to road traffic are:

- Section 41, which imposes upon the highway authority the duty to maintain the highway and
- Section 58, which provides the highway authority with the defence

  "(1) … to prove that the authority had taken such care as in all the circumstances was reasonably required to secure that the part of the highway to which the action relates was not dangerous for traffic"

  For the purposes of which defence:

  "(2) … the court shall in particular have regard to the following matters:—

  (a) the character of the highway, and the traffic which was reasonably to be expected to use it;

  (b) the standard of maintenance appropriate for a highway of that character and used by such traffic;

  (c) the state of repair in which a reasonable person would have expected to find the highway;

  (d) whether the highway authority knew, or could reasonably have been expected to know, that the condition of the part of the highway to which the action relates was likely to cause danger to users of the highway;

> *(e) where the highway authority could not reasonably have been expected to repair that part of the highway before the cause of action arose, what warning notices of its condition had been displayed;*
>
> *but for the purposes of such a defence it is not relevant to prove that the highway authority had arranged for a competent person to carry out or supervise the maintenance of the part of the highway to which the action relates unless it is also proved that the authority had given him proper instructions with regard to the maintenance of the highway and that he had carried out the instructions."*

The government's view, in February 2015, of the likely impact of autonomous driving technology upon the statutory duty of highway authorities to maintain roads was that:

> *"the standard that now passes as "reasonable" may or may not be enough when fully automated vehicles are part of the equation … If there are deficiencies in that technology, it is possible that highway authorities would need to update and improve their standard practices (for example in relation to the maintenance of road markings) in order to avoid liability in future. Alternatively, any uncertainties as to liability could be addressed through new legislation"*[2]

2 UK Department of Transport paper February 2015 - "The pathway to driverless cars: a detailed review of regulations for automated vehicle technologies" at pp.91-92, para.12.21.

The Highways Act and emerging driverless technology

Perhaps roads are not given much attention in relation to driverless technology because they seem, by contrast, rather low-tech.

If so, that is unfair. The road network is complex in its design, its structure and in its supporting communications technologies (which include a great deal of surveillance cameras, for example).

The roads themselves, however, are not "wired" to the vehicles in any controlling sense. That might be one of the reasons that they are sometimes passed over in this discussion.

There is an echo here of the anecdote about the pen developed for use in space. The story has it that NASA needed to develop a way for astronauts to write in zero gravity. The problem, it was thought, was that ink would not flow to the nib of a pen in low gravity. So a great deal of money was spent developing a pen whose mechanism would compensate for zero gravity, to allow the ink to flow.

The Russian space agency, the anecdote says, approached the problem another way, by issuing cosmonauts with a pencil.

However accurate (or not) that story is, the point is similar in relation to driverless cars and roads. Driverless cars need to be able to distinguish between different types of road user and react accordingly. Artificial intelligence is needed to do that. But a way to make the car's job easier might be to segregate the layout of the road, so as to group different road-users in different areas.

Segregation of traffic into different lanes might be considered a potential "maintenance of road markings" solution, to adopt the UK government's term from February 2015. However, it would one of a number of solutions, so it would not necessarily be a breach of duty to adopt a different solution.

Like a pencil, altering road markings to allow easier recognition by driverless cars could not be guaranteed to work on every occasion. But it would reduce the frequency of incursions and make errant users easier to spot. It might be an elegant, low-tech component of an otherwise highly sophisticated technology.

So how the factors applicable to the Section 58(2) defence in the 1980 Act might apply to driverless cars from 2017 onwards is not clear, though the factors in Section 58(2) could lend themselves to that situation.

The wider law of Infrastructure

Highway maintenance is only part of the infrastructure equation. If reliable network coverage is a prerequisite of driverless cars, because they need to communicate with each other, then the country needs reliable network coverage, nationwide.

As set out in the Introduction, the many environmental law implications of driverless technology are beyond the scope of this book. For example, driverless cars are only one part of an expanding transport technology.

The implications of transport development as a whole affect the law relating to land ownership, land use and planning law. High policies as to energy generation and pollution control looms overhead.

Efficient traffic control is a key tool of all of those policies. The duty of a traffic authority to control traffic efficiently is set out in Section 16 of the Traffic Management Act 2004 -

*"**16 The network management duty***

(1) It is the duty of a local traffic authority [or a strategic highways company ("the network management authority")] to manage their road network with a view to achieving, so far as may be reasonably practicable having regard to their other obligations, policies and objectives, the following objectives—

(a) securing the expeditious movement of traffic on the authority's road network; and

(b) facilitating the expeditious movement of traffic on road networks for which another authority is the traffic authority.

(2) The action which the authority may take in performing that duty includes, in particular, any action which they consider will contribute to securing—

(a) the more efficient use of their road network; or

(b) the avoidance, elimination or reduction of road congestion or other disruption to the movement of traffic on their road network or a road network for which another authority is the traffic authority;

and may involve the exercise of any power to regulate or coordinate the uses made of any road (or part of a road) in the road network (whether or not the power was conferred on them in their capacity as a traffic authority)."

– and that is a duty which driverless car technology seems ready to assist, as well as to push into new territory.

SUMMARY

As with data, driverless car technology will affect our roads and our local environments in far-reaching ways. The potential for this new technology to affect safety, for example, beneficially in a number of ways (by injury and pollution-reduction) is striking.

CHAPTER EIGHT
INSURANCE

INTRODUCTION

As described in the "Near Future Accidents" chapter, the government's January 2017 response to the consultation in relation to driverless cars proposes the extension of compulsory insurance to semi-autonomous cars being "driven" in automated mode, and the retention of the EU right to issue a claim directly against a motor insurer[1].

This chapter discusses further insurance risks, in the likely near-future of a "mixed fleet" of partially autonomous and old-technology vehicles, arising from:

- Employers' liabilities
- Data storage
- Malfunctions
- Crime

The objective of this chapter is to broaden the view of potential insurance risks, in the transitional period before fully driverless technology becomes dominant.

1 Dept of Transport's "Pathway to Driverless Cars", government response to consultation, January 2017. The insurance proposal also identifies limited exceptions to the insurer's cover, namely that the crash was caused by (i) the driver making unauthorised alterations to the car's operating system or (ii) failing to update the car's software. Hacking by a third party would not exclude cover. See page 11, 3.13.

Employers' Liabilities

As discussed in the "Accidents" chapters, the Employers' Liability (Compulsory Insurance) Act 1969 requires an employer to provide insurance "*against liability for bodily injury or disease sustained by his employees, and arising out of and in the course of their employment in Great Britain in that business*".

One of the key opportunities of autonomous vehicle technology is (as the British government's Department of Transport put it, in February 2015) to release time spent concentrating on the road -

> *"allowing drivers to safely use the journey time however they wish, from reading a book, to surfing the web, watching a film, or just chatting face to face with other passengers."*

Those are recreational activities. But time released by automation would also allow an employee to carry out other tasks than driving – eg. checking lists of orders, responding to emails etc. If semi-autonomous vehicles were deemed safe for the recreational purposes, then it is likely that employers would follow suit. Some former "driving" jobs might then evolve into jobs mainly concerned with clerical tasks – but in a delivery vehicle that functioned mainly as a mobile office.

In that situation, the employer's common law duties of care (particularly to provide safe systems of work, safe work equipment, and a safe place of work) would probably apply.

And breaches of those duties resulting in injury (for example, failure properly to train an employee in the use of the emergency halt control on the vehicle whilst carrying out other tasks) might trigger liabilities, which would in turn engage the employer's compulsory insurance.

Data Risks

As discussed in the "Data" chapters, the risks of physical collisions might pale in comparison to the risks of data leaks – as autonomous technology improves safety but, while doing so, processes and stores more and more information.

The financial risk of data leaks – particularly of the loss of banking information – might become more prominent in public debate around driverless cars. Insurance against such risks will follow.

Malfunctions

A car malfunction, in the age of autonomous vehicles, is more likely to be a malfunction in a network, affecting many vehicles, than restricted to one car.

The financial damage of delays in road transport might be more akin to that of flight delays. And so, again, future car insurance (or, more aptly, transport insurance) might look more like a current travel insurance policy.

Crime

Theft from an automated vehicle might more profitably be theft of information (particularly of financial details) than of physical items. That might particularly be so because of the detectability (by cameras on the car) of criminals physically present by the car.

So – like many frauds now – prevention might become mainly a matter of digital security. And theft insurance (as, eg., in car hire agreements) provided as an extra contractual charge, on top of the bill for transport.

SUMMARY

The motor insurance policy could evolve, during the transitional period, from an organism focussed entirely upon physical collisions, into a creature with many different concerns.

Among those concerns will be the effects not just of collisions but also of data leaks, malfunctions and criminal actions.

For consumers, the motor insurance policy of the future might look more like a present-day travel insurance policy (encompassing a multitude of risks) than a traditional motor policy.

If vehicle ownership dwindles, due to the greater ease of app cars, then a consumer motor insurance policy might even be renamed to reflect its new character – as a *transport* policy, for first party loss.

Transport insurance might become a more commercial field – with insurance of transport industries (of vehicles transporting employees and paying passengers) remaining compulsory. Some companies might self-insure.

In the more distant future – when fully driverless vehicles dominate – insurance taken out by the individual might complete its evolution into a transport policy for first-party loss only (there being no liability that a passenger in a driverless vehicle might owe to a third party; all liability would be upon the operator and manufacturer of that vehicle).

And the longevity of a transport policy might differ as well. Infrequent passengers, not owning their own car, might prefer to pay for their first-party insurance, on a "pay as you go" basis, signing up for insurance digitally at the start of each journey and paying at the end – the charge calculated according to the journey length (and perhaps also including different premiums, according to the areas travelled).

PART THREE

THE FULLY AUTOMATED FUTURE

CHAPTER NINE
FUTURE DATA

INTRODUCTION

This chapter speculates as to the law of data protection in a possible far future, in which driverless vehicles have become the dominant transport technology, as well as one of the many technologies that process personal data.

It speculates as to the ways in which society might deal with automated storage and movement of data in a world in which data is not just stored on computers but is used by artificially intelligent, networked devices, to make decisions on our behalf.

In that imagined, future world, both the scale and the nature of automated use of data will have expanded far beyond the use of information today.

As in all of the chapters in Part Three of the Book, this is a truly speculative exercise. The form of future technologies – the facts will that will direct the law cannot be predicted.

This chapter considers some possibilities, however, in relation to three areas of potential concern:

- Expanding data risks
- Security measures, and
- Consent to use of data

FUTURE DATA RISKS

Although we are concerned, in the present day, with the security of our data, we are also generally prepared to offer personal information to automated systems, in order to enjoy a service.

Internet shopping is an example: the ease of ordering goods online, from home or mobile devices, and having them delivered to that address, tend to outweigh the suspicion that a customer might have as to the insecurity of that transaction.

By contrast, our instincts generally tend against agreeing to give a person our bank details.

The advantages of the technology have changed our behaviour.

We cannot be certain of this, but it is fair to speculate that the same might be true of driverless vehicles. Now, we are used to seeing a human driver in any vehicle that we enter. Many of us drive ourselves. We are conditioned to place greater trust in a human driver than in an automated system.

That is the case, even though we know that human drivers can be flawed in their decision-making, and even dangerous. The occurrence of road traffic accidents, and the frequency of such accidents, tell us so.

It is a reasonable prediction that, in a further future, we might come to trust fully-automated vehicles with our physical safety. If so, that should lead to driverless vehicles being the dominant road transport technology.

And if that happens, it will probably be because the driverless vehicles are networked and able to process large amounts of information.

Some of that information (eg. a person's destination or payment details) will be "personal information", both as we now understand it and – probably – as we would commonly understand it in that future.

Should driverless technology become widespread, it is likely that its processing of personal information will go beyond personal information as to destination and payment. A driverless vehicle would probably – as a commercial proposition – offer its passenger entertainment during the journey, including online browsing and shopping (the networked equivalent of "in-flight" entertainment and sales).

So the scale of personal information processed by driverless vehicles would be at least as much as current mobile communication devices such as smartphones. And the likely improvements in technology suggest that the amount of personal information processed would be greater than current levels.

FUTURE DATA SECURITY

With an increase in the scale of personal information offered by customers to automated systems, data security will become more important. That might be reflected in the following developments:

- A trend towards more widespread and improved anonymization of personal data. This would be moved primarily by technology but there would be powerful commercial motives to allow this (to increase commercial revenue by making customers feel safer in their shopping). There would also be strong legal reasons to do so: reliable anonymization software, for example, could reduce the need to regulate in ever-increasing detail as to the permissible circumstances of personal data storage. Over-complex regulation becomes difficult to enforce and is a disincentive to investment in new technologies.

- Increased investment by states in data security. The phenomenon of the hostile hack of important computer systems – whether by a hostile state or by individuals – was in my teenage years a subject for Hollywood. It is now frequent and real. It is a reported feature of industrial espionage and even of elections. As infrastructure includes a larger virtual element, data security will increasingly become part of national defence.

- Increased investment by individuals in data security. Anti-malware etc software is common. But individuals' vulnerability to what we now call cyber-crime will, sadly, inevitably increase with the expansion of mobile data, however advanced the security technologies become. Driverless cars will expand mobile data use considerably. The form of future crime is unpredictable but it is bound to innovate. The market for effective cyber-security products is likely to increase, as is the market for:

- Cyber Insurance, as the last line of financial protection in the event of a security breach. In line with current motor insurance providing cover for (broadly) accidental as well as criminal events, cyber-insurance is likely to cover accidental and criminal damage resulting from loss or misuse of personal data. Whether such insurance takes the form of an annual "Data Coverage Policy", covering data loss in any situation (eg. whether in a driverless app car or shopping online), or is a risk covered under another type of policy ("Driver's insurance", if the person continues to drive his own vehicle, for example, or under a Household or Travel insurance policy) remains to be seen.

FUTURE CONSENT TO USE OF DATA

As discussed in Chapter Four, the EU Working Party envisages consents becoming more specific, in order to make clear precisely which pieces of data the user is happy to be released to the commercial or other concern asking for it.

That aim is combined, for good reasons, with the requirement that the request for consent be intelligible. So the consent request needs to be precise and intelligible.

The App (say) asking for the consent will also need to be alive to the inferences that might be drawn from the information that it is asking for. For example, a person asking a driverless app car to take him to a place of worship might be offering information which, if stored, would be in a category of special sensitivity and protection. If so, only an explicit consent to such storage, in response to a clear and comprehensive request, might suffice to satisfy the App business's legal obligations and the user's legal rights.

For the reason set out at the end of Chapter Four, the sort of reasoning required of the artificial intelligence operating the App might in that case be achievable, but more complex matters of inference might be perilous. Anonymisation software, or software that somehow prevents storage of the destination data, might solve the problem.

But the issue of data consents emphasises the complexity of simple requests, and the inferences that might be drawn.

SUMMARY

Current data protection law is replete with detail and guidance. Current data laws can be infringed – potentially with criminal consequences – indirectly, by unjustified storage of information that *implies* protected data.

This presents technological, social and legal challenges: chiefly as to whether equipment, people and the law can agree as to the acceptable balance of privacy on the one hand and effective transport services on the other.

CHAPTER TEN
FUTURE RISKS & INSURANCE

INTRODUCTION

Accidents & Crimes

The Oxford English dictionary defines "accident" as "an unfortunate incident that happens unexpectedly and unintentionally".

Now, a road traffic accident implies a collision involving a vehicle, physical damage to property (including the vehicle) and personal injury to its occupants and bystanders. Inappropriate speed and insufficient control of the vehicle are frequent causes of accidents. Road traffic accidents occur much too often and have devastating effects. Their reduction is a major spur to driverless car technology.

In the fully-driverless future, the greater number of unintended and unwelcome events might not be collisions but crimes – chiefly hacking of car systems for information.

Collisions, as they become a less common event, will provide a lesser disguise for fraud, so road traffic collision fraud will diminish. Road traffic information fraud will probably, however, increase.

So prevention of data fraud will be a major area for transport security. The risks of data theft and fraud will become a major subject for insurers.

Infrastructure will be required to cater for:

- Personal security – by segregation of road users, to increase safety against collisions, and by planning of space, lighting etc to reduce risk of robbery

- Data security – by a robust and secure system of data networks

As to collisions, in this imagined future of fully automated vehicles, control of vehicles is automated, speed is constantly regulated at an appropriately safe level and vehicles communicate with each other in order to move safely together.

Manually controlled vehicles are obsolete. Human distraction is removed from driving and roads are designed to segregate their users by category (pedestrians, cyclists, automated vehicles).

The roads in urban areas are populated far less by private vehicles and are dominated by app cars. Parked underground, those cars emerge when summoned by a customer's smartphone or watch. They are much smaller and lighter than today's cars, requiring neither the space for a driver or controls, nor a large engine. They make little noise. Their electrical engines do not pollute.

On the motorways, individual cars are relatively few in number. The greater part of motorway traffic consists of driverless buses and lorries, "platooning" in groups of three vehicles driving close together to save energy, in prescribed lanes.

Speed is automated. Collisions causing injury and damage are very rare.

The need for compulsory insurance for the damage and injury caused by accidental collisions diminishes. The case for such compulsory insurance weakens. Insurance reforms[1].

1 The speculations in this chapter are mine, but the call for reform of compulsory motor insurance has been made by those far more expert: see the DoT January 2017 "Pathway to Driverless Cars" consultation response paper at p.11, para.3.35 (describing the submissions of Professor Robert Merkin QC, Professor of Commercial Law, University of Exeter, and of the Insurance Law Research Group, University of Southampton).

What "accidents" might happen in that future?

This chapter speculates as to:

- the accidents that road traffic might produce in that future, and
- the laws that might evolve to govern them.

The facts set out in this chapter are, inevitably, entirely speculative. But the version of the law set out – though it springs from that imaginative scenario follows lines of established principle. Something wholly different could happen.

FUTURE ROAD TRAFFIC ACCIDENTS

Road traffic accidents will typically not, mercifully, involve physical collisions between road-users.

Collisions

When collisions involving self-propelled vehicles do occur, they are aberrations, due mainly to systemic failures. They are diagnosable. "Black box" type of equipment on the vehicles record data from many sources and reconstruct the circumstances of any collision. The absence of a human driver removes the need to draw inferences as to physical incapacity, emotion and motive. The movements of the vehicle are clear.

The law might evolve to assign fault for these rare occurrences on a strict liability basis: assigning responsibility for any injured party's claim to the body that economically profits from the faulty vehicle, on the same logic as vicarious liability for employers.

Contributions from the producers of any allegedly faulty component of that vehicle are left to the responsible commercial organisations to agree between themselves. With the cooperation of insurers, the transport industry adopts its own court of arbitration for such disputes.

Insurance relating to vehicle collisions is provided to passengers by the manufacturers and transport companies, on a compulsory public liability basis for those businesses.

Road collision insurance remains compulsory for individuals only for the relatively few consumers who own and permanently keep their own vehicles. The need to own an individual vehicle diminishes as the costs favour the use of app cars. Road traffic changes from a community of drivers to one of passengers.

With the reduction in collisions, the risks and therefore the premiums for road collision insurance reduce.

Information

Data loss and theft increase. Transport companies' major risk is data security.

The major physical safety concern arises from hacking of systems and from individual crime. Attempted theft of the valuable physical components and data stored in app cars increases.

Insurers see a rise in data security and theft cover. As the number of road traffic collisions and consequent injuries and deaths fall, data and personal security of app car passengers become the dominant risks relating to road transport.

Optional "personal transport insurance", including data security, personal security and collision cover, replaces individual, compulsory road traffic insurance.

SUMMARY

The changes in road vehicle technology that might occur over the coming decades could be radical. They should be highly beneficial in terms of personal safety.

They might (as I speculate) shift the risks of road transport away from physical collisions involving motor vehicles.

The new areas of risk might equally arise from the technology – particularly because of the amount of data that the new vehicles are likely to process, some of which will be attractive to criminals. As transport joins "the internet of things", future transport crime might become increasingly virtual.

Some risks will remain, even in this less risky world. But old and new technology – careful road planning in particular – will combine, to reduce the risks of travelling our streets.

CHAPTER ELEVEN
FUTURE ROADS

INTRODUCTION

As mentioned in Chapter 7, the transport visionaries of 60 years ago imagined roads that would propel cars to their destination without any need for human control[1].

How might our highways and streets appear in future decades, if driverless cars become the dominant technology?

How many driverless CARS will there be?

Dr Robert E Smith, of University College, London, posed this question at a lecture in July 2016.

His point related to energy-efficiency. Would the economics of energy generation favour –

- small vehicles, each powered by its own engine,

  or

- vehicles capable of carrying larger numbers of people than a single car, but still powered by a single engine (so more energy-efficient)?

In Dr Smith's phrase, should a driverless car actually be a tram?

The point is interesting and the outcome unpredictable. The advertising that accompanied the dream of an electronic highway in the 1950's showed a family, in an automated car. A photograph of my great grandfather and children (including my grandmother) in their car in the 1920's is a photograph of a family unit, enjoying a new technology.

1 Chapter 7 ("First there were electronic highways") of "Driverless" Intelligent Cars and the Road Ahead" by Hod Lipson and Melba Kurman (MIT Press, 2016).

We associate the car with freedom of movement for small units of people – an individual, a couple or a family. The attraction of the mass-produced car in its infancy was the promise of independence of movement, away (far away) from the crowd.

But the modern challenges are different: the negatives of traffic congestion, the dangers of collisions and of air pollution reaching startling levels.

Smog has returned to British cities through a different cause – traffic exhaust rather than coal fumes. And in cities in the developing world, smog is – perversely – a sign of greater affluence and of vehicle ownership.

So Dr Smith's question boils down to this – will "driverless" technology lead mainly to driverless cars, or to driverless *transport*?

Will driverless cars be less or more attractive objects to own? Will we choose to hire them at will – essentially as passengers in an automated taxi? Will driverless road technology translate equally or to a greater extent into buses and trams?

The answers to those questions will tell us how our streets will look in future decades. Will there be as many cars parked on our streets? Will we all feel a need to own a car, or might that desire slip away, if our perception of public transport were to improve?

SUMMARY

Road design and virtual infrastructure will be crucial to the success of driverless technology.

Road design raises questions as to environmental matters – including questions relating to the organisation of space and energy generation. Whether road transport will continue to be dominated by small vehicles is a key issue, affecting those and other matters.

The answers to those questions also depend upon decisions we take now – particularly in relation to the road and connectivity infrastructure that driverless technology will require.

CONCLUSION

Writing about driverless cars seems, from one perspective, a rather joyless subject.

My Dad's father worked in the motor factories of the West Midlands. When he retired, he had a job delivering cars to London. That was during my childhood, when we lived in a cul-de-sac close to the M1. So my grandpa would pop in to see us – once, I remember, driving a Ferrari (bright red, of course).

My grandpa was not one for rules. He taught my Dad (still an excellent driver and car enthusiast) to drive when he was only fourteen, somewhere in Birmingham – and presumably beyond the gaze of his policeman brother-in-law.

My grandpa regarded the introduction of the mandatory seatbelt law as the worst incursion on his civil rights. He would carefully click the driver's seatbelt home, then sit on top of it.

So it's with some sense of disapproval that I've written a book about the law of driverless cars, enthusiastically. Driving a car is often a joy.

And while my grandpa admired my bike when I was eight, I'm pretty sure that he would have disapproved of my still cycling in my mid forties (a different bike).

If cars are about to enter the next stage of their evolution, our world will change. My grandparents' home city, largely destroyed by war, was rebuilt for a populace whose transport would be personal, accessible and revolutionary. The car has shaped our cities, forged great paths across the country and has changed the world.

So its next mutation will reshape everything again. Proof of that lies in the difficulty of thinking about the world beyond. How might we behave? What laws will we need?

Those might prove to be the smaller questions. It could be a safer, quieter world.

'Baby Steps: A Guide to Maternity Leave and Maternity Pay' by Leah Waller
'The Queen's Counsel Lawyer's Omnibus: 20 Years of Cartoons from the Times 1993-2013' by Alex Steuart Williams

27183718R00088

Printed in Great Britain
by Amazon